21st CENTURY

LIFESKILLS

MATHEMATICS

Budgeting & Banking Math

SADDLEBACK
EDUCATIONAL PUBLISHING
www.sdlback.com

ISBN-13: 978-1-61651-407-5
ISBN-10: 1-61651-407-8
eBook: 978-1-61247-001-6

Printed in the United States of America

18 17 16 15 14   5 6 7 8 9

Table of Contents

Unit 5: Rental Budgeting

Unit 6: Self-Employment

Unit 7: Going on a Trip

Unit 8: Keeping the Books

To the Student

Welcome to *Budgeting & Banking!* This is Book 4 of the *21st Century Lifeskills Mathematics* series.

The goal of this book and the other books in this series is to build your confidence and practical math skills. You will use these math skills in everyday situations throughout your life.

You solve problems and make mathematical decisions every day. You compare products and make choices about what to buy. You work to earn money. You decide what to spend and how much to save.

21st Century Lifeskills Mathematics gives you strategies to solve everyday math problems in a variety of ways. It strengthens your skills and gives you practice with many different math topics. Each of the six books presents topics you are likely to encounter in everyday life. Each book includes problems that involve estimation, equations, mental math, calculators, and critical thinking. Each book includes additional topic-specific skills such as graphing, averages, statistics, ratios, and measurement.

Each unit begins with a preview lesson, which models and explains the types of problems you will encounter in the unit. Then there are five lessons, at least one of which is usually a game. Each unit ends with a review of the unit concepts. There are illustrations and graphic art throughout.

Here is a list of the titles of the other books in the *21st Century Lifeskills Mathematics* series:

Book 1: Everyday Life
Book 2: Home & School
Book 3: On the Job
Book 5: Smart Shopping
Book 6: Sports, Hobbies, & Recreation

With review and practice, you will build your math skills and learn to approach everyday mathematical situations with confidence! *21st Century Lifeskills Mathematics* will help you become a successful problem solver!

Unit 1 — Paying Your Way

Preview

How You Will Use This Unit

Paying your way involves many different things. Sharing day-to-day costs and budgeting for special items are just two examples. You probably also consider saving some of the money you make. You may also want to compare services from banks. As you compare options and make choices, you will often use math. The math skills you use include mental math and estimation, basic operations and equations, statistics, and ratios and proportions.

What You Will Do in This Unit

In this unit, math steps demonstrate how to solve problems. These steps can help you answer questions such as these:

You share an apartment with a friend, and share equally the cost of rent, utilities, and food. How much money do you budget for these expenses each month?

Out of your weekly income, you budget $20 toward new windsurfing equipment. How much do you have toward your new equipment after six weeks?

You keep $1,000 in your checking account. When this grows to $1,500, you transfer $500 to savings. What is the percent increase in your checking account just before you transfer funds?

Your bank offers free checking when you maintain a minimum daily balance of $2,000. When you do not maintain this balance, the bank charges a $5 monthly service charge, and a $0.05 transaction charge. What is the fee if your daily balance is $800, and you write 15 checks?

What You Can Learn from This Unit

When you complete this unit, you will have used mathematics to work problems related to paying your way. These problems are similar to those that may actually occur in your daily life.

Lesson 1 — Sharing Day-to-Day Costs

Example Milan and Dale share a student apartment. They share equally the cost of rent, utilities, food, and miscellaneous living expenses. Milan draws a circle graph to show their total costs for this month.

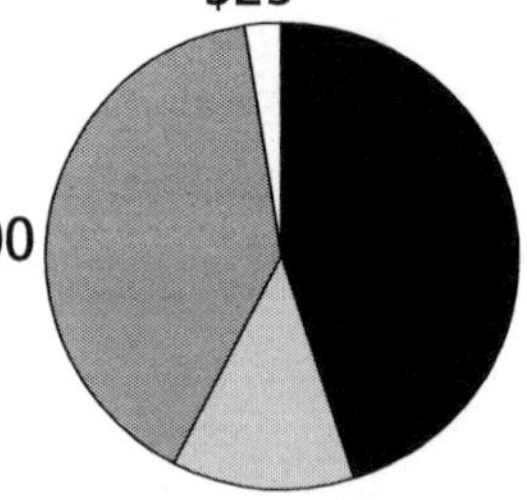

a. What fraction of the total costs is the rent?

b. What dollar amount does each student budget for these expenses per month?

Solve

Step 1: First, find the total expenses by adding.

$\$450 + \$125 + \$400 + \$25 = \$1000$

Step 2: Then, write the rent as a fraction of the total. Simplify the fraction by dividing out the common factors.

$\frac{\$450}{\$1000} = \frac{9}{20}$

Step 3: Underline the sentence that tells how Milan and Dale handle apartment costs.

<u>They share equally the cost of rent, utilities, food, and miscellaneous living expenses.</u>

Step 4: Divide to find x, the amount of the two equal shares of the costs.

$$\begin{aligned} x &= \text{total cost} \div 2 \\ &= \$1,000 \div 2 \\ &= \$500 \end{aligned}$$

Answer the Question

Step 5: **a.** The rent is $\frac{9}{20}$ of the total cost.
b. Each student budgets $500 each month.

Now try these problems.

1. Milan and Dale share equally the costs of rent, utilities, food, and other living expenses equally. Their rent increases to $549. They reduce their other shared costs to $366.

   a. What fraction of the total is this increased rent?

b. What amount does each student budget for these expenses per month?

Answer: a. The rent is ______ of the total.

b. Each student budgets $__________.

2. Melissa pays a car insurance premium of $160 each month. She starts to carpool to work. She informs her car insurance company of the change. The company takes 5% off her car insurance premium. How much is her monthly car insurance premium now? Circle the bills and write the number of those bills under each, to represent this amount.

_______ _______ _______ _______ _______

3. Jo-emma rents a house. She makes an agreement with the owner about the garden. The owner buys the plants, and Jo-emma plants and waters them. The owner saves $100 a month for a gardener. The owner reduces Jo-emma's rent each month by this amount. This represents a 20% savings each month on Jo-emma's rent. What is the amount Jo-emma actually pays each month for rent?

A $160 **B** $200 **C** $400 **D** $500

4. Ruben budgets $20 a week for lunch. On Monday, Alex and Melvyn invite him to go to lunch with them. Ruben's lunch costs $5.75. Alex's lunch costs $6.25. Melvyn's lunch costs $6.00. They split the total bill equally between them. How much does Ruben have left for lunches for the rest of the week?

Answer: $__________

☆ *Challenge Problem*

You may want to talk this one over with a partner.

Can you use the same sequence of symbols, in the same order, to make each equation correct? Choose from these symbols: +, –, ×, ÷, (). Write an equation to support your answer.

4 __4__4__4__4 = 1 5__5__5__5__5 = 1 6__6__6__6__6 = 1

Lesson 2 → Budgeting for Special Items

Example Out of her weekly income of $480, Carmen budgets $20 toward new windsurfing equipment.

a. What percent of her weekly income is the amount she budgets?

b. She plans to continue to budget this amount each week. How much will she have toward her windsurfing equipment after six weeks?

Solve

Step 1: Underline the sentence that tells the amount Carmen budgets of her income.

Out of her weekly income of $480, Carmen budgets $20 ...

Step 2: Write an equation that gives the budgeted amount as a percent of her income.

$$\text{Percent} = \frac{20}{480} \times 100\%$$
$$= 4.167\%$$

Step 3: Use a proportion to find the amount x that Carmen has after six weeks.

First, use words to write the proportion.

$$\frac{\text{amount in one week}}{\text{one week}} = \frac{\text{amount after six weeks}}{\text{six weeks}}$$

Second, write the proportion using numbers.

$$\frac{20}{1} = \frac{x}{6}$$

Step 4: Solve for x.

$$x = 20 \times 6$$
$$x = 120$$

Answer the Question

Step 5: **a.** Carmen budgets 4.167% of her weekly income.
b. After six weeks, she has $120 toward her windsurfing equipment.

Now try these problems.

1. Out of his monthly income of $1,200, Chaz budgets $60 for a digital camera.

a. Chaz budgets ______% of his monthly income for a digital camera.

b. He plans to continue to budget the same amount each month towards the camera. Which equation gives the amount he will have after six months?

A $\frac{x}{60} = \frac{1}{6}$ **B** $\frac{x}{6} = \frac{60}{1}$ **C** $\frac{x}{1} = \frac{6}{120}$ **D** $\frac{x}{6} = \frac{1}{1,200}$

2. As a book illustrator, Ravenna takes about four hours to create one finished drawing. She likes to create at least 15 drawings per week. Ravenna budgets her time so that she has Sundays off. This week she completes four drawings on Monday and two on Tuesday. How can she budget her time for the rest of the week and meet her goals? Fill in a number of working hours for each day.

Monday	Tuesday	Wednesday	Thursday	Friday	Saturday	Sunday
		____ hours	____ hours	____ hours	____ hours	____ hours

3. Veronica budgets $19.95 a month for a new E-zine she reads. Her monthly expenses total $760 per month before this new expense. Her income is $800 per month. How much money will she have left over each month after this new budget item?

Answer: $__________

4. Dimitri buys a car for $3,600. He agrees to pay for it in equal monthly installments over two years. His car insurance is $1,600 per year. He pays it in equal monthly installments. Write a set of steps to calculate the change in Dimitri's monthly expense budget.

Answer: ______________________________________

☆ *Challenge Problem*

You may want to talk this one over with a partner.

Ask a friend to pick (without telling you) the number of times a week that they would like to have dinner out (more than once, less than 10). Then give them these directions: Multiply this number by 2. Add 5. Multiply by 50. If they have already had their birthday this year, add 1753. If they haven't, add 1752. Add 1 if this year is 2004. Add 2 if this year is 2005. Add 3 if this year is 2006 … and so on. Now subtract the four digit year that they were born. Ask your friend what the final three-digit number is.

You can decode this final number! The first digit is the number of times your friend chose to eat out. The next two digits is the age of your friend! See if you can figure out why this always works.

Lesson 3

Saving, a Little at a Time

Example Roe keeps $1,000 in her checking account. When this grows to $1,500, she transfers $500 to her savings account.

a. What is the percent increase in her checking account just before she transfers funds out?

Roe keeps $3,000 in her savings account. When this account has the same percent increase, she transfers the excess to an investment account.

b. What amount does her savings account reach just before she transfers out funds?

c. How much money does she transfer out?

Solve

Step 1: Underline the sentence that tells how much Roe keeps in her checking account.

Roe keeps $1,000 in her checking account.

Step 2: To find the percent increase in Roe's checking account, first subtract to find the amount of the increase.

$1,500 – $1,000 = $500

Percent of change (increase or decrease) equals $\frac{\text{amount of change}}{\text{original amount}}$

Step 3: Next, find the percent by comparing the amount of change to the original amount.

$\frac{50}{1,000} \times 100\% = 50\%$

Step 4: Now, use the same percent to calculate the amount of increase for her savings account.

50% of $3,000 = $1,500	Find 50% of the amount she keeps in that account.
$3,000 + $1,500 = $4,500	Add this increase.

Answer the Question

Step 5: **a.** Roe's checking account has increased by 50% when she transfers out funds.

b. Roe's savings account grows to $4,500.

c. Then she transfers out $1,500.

Now try these problems.

1. Monty keeps $800 in his checking account. When this grows to $1,000, he transfers $200 to his savings account.

a. When Monty's checking account increases by ______%, he transfers funds out.

Monty keeps $2,000 in his savings account. When this grows by the same percent increase, he transfers the excess to an investment account.

b. What amount is in his savings account when he transfers out funds?

c. How much money does he transfer out?

Answer: Monty's savings account grows to $__________.

Then he transfers $__________ out.

2. To save money, Donna separates her essential expenses from her optional expenses. On her telephone bill, the essential service charge expense is about $20.00 a month. Her optional long-distance calls average about $50 per month. She reduces her optional expenses by 20%. How much does she save per month?

A $5 **B** $10 **C** $20 **D** $50

3. During the first year of their marriage, Dawn and Jeff save $1,000. With $600, they buy a three-year Certificate of Deposit (CD). With the remaining money they buy a used sewing machine and a chain saw. Dawn uses the sewing machine to make gifts. She spends $300 less on buying gifts. Jeff uses the chain saw to cut up logs. He spends $250 less on their heating bill. Circle the numbers that represent savings. Draw squares around the numbers that represent expenses.

$600 $400 $300 $250

4. Eduardo earns $25,000 a year this year. He estimates that he will get an 8% increase in salary each year. He plans to save 10% of his income each year. Complete the chart. How much can he save in 3 years?

Year	Income	Year's Savings	Accumulated Savings
1	$25,000	10% of $25,000 = $	$
2	1.08 × $25,000 = $	$	$
3		$	$

☆ *Challenge Problem*

You may want to talk this one over with a partner.

You count the change you have been saving. You have exactly enough for the hubcaps that cost $98. You have four five-dollar bills, 21 one-dollar bills, 141 quarters, 69 dimes, and a lot of nickels.
How many nickels do you have? **Answer:** __________

Lesson 4 — *Using Bank Services*

Example Marta's bank offers free checking when she maintains a minimum daily balance of $2,000. When she does *not* maintain a minimum daily balance of $2,000, the bank charges a fee. The fee includes a monthly service charge of $3, and a 10-cent charge for each check or withdrawal transaction. Marta does *not* maintain a daily balance of $2,000. Write an equation for the bank fee for one month.

Solve

Step 1: Use f for the bank fee for one month.

Use t for the number of check or withdrawal transactions in one month.

Write an expression for the charge for t transactions.

Charge for t transactions = $\$0.10 \times t$ 10 cents for each transaction.

Step 2: Now, write an equation for the bank fee, f, for the month.

$f = (\$0.10 \times t) + \$3$ Add the $3 monthly service charge.

Answer the Question

Step 3: The equation for the bank fee for one month is: $f = (\$0.10 \times t) + \$3$.

Now try these problems.

1. Cody's bank offers free checking when he maintains a minimum daily balance of $2,500. When he does *not* maintain this minimum daily balance, the bank charges a fee. The fee includes a $4 monthly service charge and a $0.05 transaction charge. Cody maintains a daily balance of $2,000. Which equation gives Cody's bank fee for one month?

   A $f = \$0$
   B $f = (\$0.04 \times t) + \$5$
   C $f = (\$0.05 \times t) + \$4$
   D $f = \$4$

2. Jere does *not* maintain the required $1,500 minimum daily balance at her bank. This month, she writes 21 checks, and makes two withdrawal transactions. Her bank charges her a

monthly charge of $3.50, and 10 cents per transaction. The bank across the street charges a flat fee of $5 per month, with no restrictions. Which bank suits Jere's banking needs better? Why?

Answer: ____

____

Deposits: $1,312.50 Checks paid: $27.97, $52.70, $345.59 Withdrawals: $200, $200, $300 Current balance: $1,483.82

3. Fia's bank offers Internet banking services. She checks the status of her checking account on-line. Her rent check for $525, that she just mailed, has not cleared yet. Will she have enough in her account to pay her $745 credit card bill in full, after the rent check clears? Write an equation and find the amount in her account if she pays these two bills.

Answer: ______ (Yes/No) ________________________ = $__________

4. Here are the yearly rates for a $1,000 Certificate of Deposit (CD) at Elton's bank:

Interest earned on a $1,000 CD	180 days: 0.65% interest	1 year: 0.75% interest

Elton considers withdrawing his $1,000 CD after 180 days. How much *more* interest can Elton earn if he leaves the CD in the bank for one year?

$1 $3.25 $4.25 $7.00

☆ *Challenge Problem*

You may want to talk this one over with a partner.

You have a bank debit card and a bank credit card. There is no charge for the debit card. But you can *not* exceed the $2,000 that you have in your checking account without incurring a charge. There is a 16% interest charge on any unpaid balance in your credit card account. But you have a $5,000 credit limit with this card. You have an outstanding balance of $2,200 against this card. You want to purchase a $1,500 sofa for your apartment. Which card do you use? Why?

____

____

____

____

Lesson 5 — It All Adds Up!

A Jigsaw Puzzle Game (for One or More Players)

The goal of this game is to build a three-by-three square. First, you look at the nine pieces. Then, you put them together so that the answers on adjacent edges of the small squares are the same.

Materials

Nine small squares (on the next page)

Directions

1. Mix up the nine small square pieces and place them face up on a flat surface. Players arrange themselves so they can see all the pieces.
2. Player 1 picks two pieces where the answer to the expression on one side of a square is equal to the answer to the expression on one side of the other square. Player 1 places the pieces so that these two matching sides are together.
3. Player 2 then takes a turn, and adds a piece to one of edges of the two pieces. (Remember that the answer on each side of the touching edges must be the same.)
4. Players take turns adding *or replacing* a piece in the puzzle.
5. The game is over when the nine pieces fit together in a three-by-three square, and the answers to the expressions on adjacent edges are equal.
6. For a more complex game, cut two sets of the small squares, mix them up, and try to build two squares at the same time. Alternatively, two teams can compete to build their own square. The winning team is the team that finishes first.

Before you play the game, try these warm-up problems.

1. Nate looks at the expressions on several of the *It All Adds Up!* squares. Which expression is equal to this expression: $1^2 + 2^2 - 4$? (Recall that any number to the zero power is equal to 1.)

   A 1 **B** $2^0 - 3$ **C** $\sqrt{4}$ **D** $\sqrt{25}$

2. When all nine pieces fit together, how many of the small squares have unmatched answers on more than one side? Explain your answer.

   Answer: ______________________________

Game Squares for *It All Adds Up!*

Copy this page and cut apart the squares to make a set of nine game pieces.

Top	Left	Right	Bottom
$\sqrt{16} + 3^2$	$\frac{1}{5}$ of 75	$\frac{1}{6}$ of 120	$70 \div \sqrt{4}$
$10 - \sqrt{1}$	0.025×120	5×9	4^2
$11 - 3$	9% of 100	$7 \times \sqrt{25}$	$1^2 + 2^2 - 4$
$3^3 - 11$	$3 \times 7 - 1$	18% of 50	$13 + 8$
3×5	$2^2 + 3$	$\sqrt{25} \times 2^2$	$2^2 \times 3$
$\frac{1}{8}$ of 88	$\frac{1}{2}$ of 10×3^2	$(3 \times 6) - 3$	$1 + 7$
6×0.5	$\sqrt{49}$	100^0	$85^0 + 10^1$
$40 - \sqrt{25}$	50% of 40	$\frac{1}{7}$ of 49	$(2^2 \times 3^2) \div 4$
20% of 60	4×5	$\frac{1}{3}$ of 9	$4^2 - 3$

Unit 1 Review

Review What You Learned

In this unit you have used mathematics to solve many problems. You have used mental math and estimation, practiced basic operations, solved equations, and used statistics, ratios, and proportions.

These two pages give you a chance to review the mathematics you used and check your skills.

✔ Check Your Skills

1. Korey and Joe share a student apartment. They share equally the cost of rent, utilities, food, and miscellaneous living expenses. Korey draws a circle graph to show their total costs for this month.

a. What fraction of the total is the rent?

b. What amount does each student budget for these expenses per month?

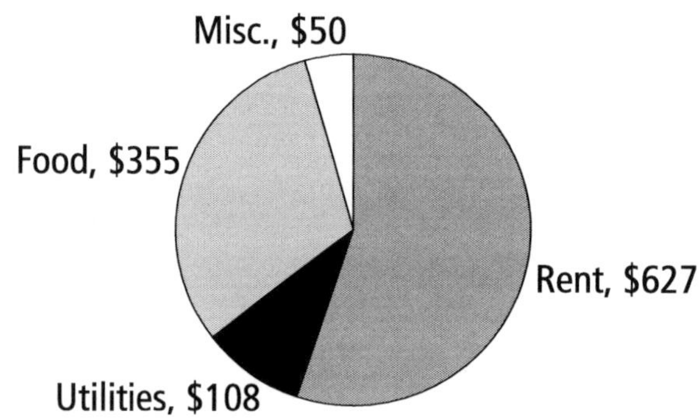

Answer: a. __________

b. __________

If you need to review, return to lesson 1 (page 2).

2. Out of his monthly income of $1,900, Seb budgets $125 for new kiteboarding equipment.

a. Seb budgets ________% of his monthly income for kiteboarding equipment.

b. He plans to continue to budget $125 each month toward this equipment. Which equation gives the amount he has after three months?

A $\frac{x}{125} = \frac{1}{3}$ **C** $\frac{x}{1} = \frac{125}{3}$

B $\frac{x}{3} = \frac{125}{1}$ **D** $\frac{x}{3} = \frac{1}{125}$

If you need to review, return to lesson 2 (page 4).

3. Arno keeps \$1,000 in his checking account.When this grows to \$2,000, he transfers \$1,000 to his savings account.

   a. What is the percent increase in his checking account just before he transfers funds out? Circle the percent inside the circle.

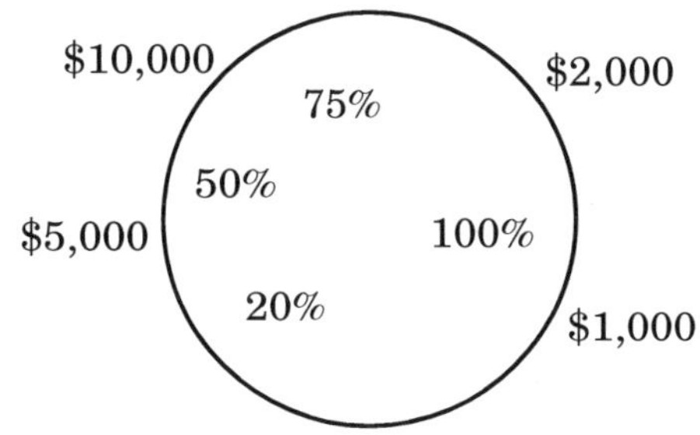

   Arno keeps \$5,000 in his savings account. When this grows by the same percent increase, he transfers the excess to an investment account.

   b. How much does his savings account grow to before he transfers funds out? Circle the dollar amount outside the circle.

   If you need to review, return to lesson 3 (page 6).

4. Brenda's bank offers free checking when she maintains a minimum daily balance of \$1,000. When she does *not* maintain this minimum daily balance, the bank charges a fee. The fee includes a \$5 monthly service charge, and a \$0.05 transaction charge. Brenda maintains a minimum daily balance of \$800. Which equation gives Brenda's bank fee for one month?

   A $f = \$5 \times t + \$0.05$ **C** $f = \$0.05 \times t + \$800$

   B $f = \$0.05 \times t + \$200$ **D** $f = \$0.05 \times t + \$5$

   If you need to review, return to lesson 4 (page 8).

Write Your Own Problem

Choose a problem you liked from this unit. Write a similar problem using a situation and related facts from your own life. With a partner, share and solve these problems together. Discuss the mathematics and compare the steps you used. If you need to, rewrite or correct the problems. Write your edited problem and the answer here.

Unit 2 Buying Large Items

Preview

How You Will Use This Unit

Buying large items involves many different things. Buying a home or a car are just two examples. You probably also consider other large expenses such as college and insurance. As you compare options and make choices, you will often use math. The math skills you use include mental math and estimation, basic operations and equations, statistics, and ratios and proportions.

What You Will Do in This Unit

In this unit, math steps demonstrate how to solve problems. These steps can help you answer questions such as these:

A car salesperson offers you two payment plans. You can pay for your car in 12 months. Or, with $1,000 down, you can pay $25 less and pay for the car in 8 months. What is the price of the car?

You want to borrow $35,000 to buy a house. For a 30-year loan at an 8% interest rate, your monthly payments will be about 0.7% of the loan. About how much are your monthly payments?

You are going to technical school. To cover your expenses, you borrow $4,000 for two years at 8.5% interest rate, compounded annually. How much will you owe at the end of the second year?

You spend $133 of your monthly income of $1,250 on health insurance. Beth spends the same fraction of her monthly income of $1,440 on health insurance. How much does she spend on health insurance each month?

What You Can Learn from This Unit

When you complete this unit, you will have used mathematics to work problems related to buying large items. These problems are similar to those that may actually occur in your daily life.

Lesson 1

Buying a Car

Example Angelina visits several car dealerships, and finally decides on which used car to buy. The salesperson offers her two 0%-interest payment plans. She writes notes of these details.

i. With no down payment, she pays for the car in twelve equal monthly installments.
ii. With a down payment of \$1,200, she pays \$40 less per month over ten months.

But what is the price of the car?

Solve

Step 1: Write an equation for the first payment plan. Use p for the price of the car. Use x for the amount she pays each month.

$12x = p$	12 months at x dollars per month gives the total.

Step 2: Now, write an equation for the second payment plan.

$\$1{,}200 + 10(x - \$40) = p$	Down payment + 10 months of (x less \$40) is the total.

Step 3: Solve this system of two equations to find the value of p.

$x = \frac{p}{12}$	Use the first equation to write x in terms of p.
$\$1{,}200 + 10(\frac{p}{12} - \$40) = p$	Substitute $\frac{p}{12}$ for x in the second equation.
$\$1{,}200 + \frac{10}{12}p - \$400 = p$	Apply the Distributive Property to multiply 10.
$\$800 = p - \frac{5}{6}p$	Combine terms and simplify.
$\$800 = \frac{1}{6}p$	Think of p as $\frac{6}{6}p$ and subtract.
$6(\$800) = 6(\frac{1}{6}p)$	Multiply each side by the denominator.
$\$4{,}800 = p$	Simplify.

Answer the Question

Step 4: The price of the car is \$4,800.

Now try these problems.

1. Calvin visits several car dealerships, and finally decides which used car to buy. The salesperson offers him two 0%-interest payment plans. He notes these details.

i. With no down payment, he pays for the car in twelve equal monthly installments.

ii. With a down payment of $750, he pays $25 less per month over ten months.

What is the price of the car?

Answer: The price of the car is $___________.

2. The motorbike that Faye likes has a price tag of $16,900. She makes a table to show various 0%-interest payment plans to pay off the bike. Which entries are not correct? Circle each incorrect entry. What error did Faye make?

0%-interest plans	Pay off in 1 year	Pay off in 2 years	Pay off in 3 years
$1,000 down payment	$1,325/month	$662.50/month	$441.67/month
$2,000 down payment	$1,241.67/month	$620.83/month	$413.89/month
$5,000 down payment	$1,158.33/month	$579.17/month	$386.11/month

Answer: ____

3. Brett looks at a new truck that costs $26,000. He reads that a new vehicle goes down in value each year by an amount equal to 17% of the original price. What is the value of the truck at the end of three years?

A $12,339 **B** $12,740 **C** $14,866 **D** $17,911

4. Danna shops for a used car. One car with basic accessories and no alarm system, costs $5,400. Exactly the same car but with all the accessories and an alarm system, costs $1,200 more. Danna's $1,600 annual insurance cost will go down by 9% with an alarm system. She plans to pay off the car in one year. Fill in the cost details for each car. Which car would you buy? Why?

Car 1: Car cost: $5,400

Insurance: $__________

Total cost: $__________

Car 2: Car cost: $__________

Insurance: $__________

Total cost: $__________

Answer: ____

☆ Challenge Problem

You may want to talk this one over with a partner.

You buy a used car for $4,300 that you will pay for by credit card. You pay $1,000 down. You plan to pay off the car in monthly installments over 3 years. The bank that issued your credit card charges an annual interest rate of 19.8%. Each month the bank charges interest on unpaid balances. You make no more purchases. How much interest do you pay at the end of the first month? $__________

Lesson 2

Buying a Home

Example Tandy thinks about buying a house that is valued at $75,000. He can make a $10,000 down payment. He will need to borrow the rest from a mortgage company. His uncle gives him a good rule to use for a rough approximation of his monthly payments. For a 30-year loan at an 8% interest rate, his monthly payments will be about 0.7% of the amount he borrows. What is an approximation of Tandy's monthly payments by this rule?

Solve

Step 1: Compute the amount that Tandy will owe on a $75,000 house, after he puts $10,000 down.

$75,000 – $10,000 = $65,000

Step 2: Now, use his uncle's rule to find an approximation for one monthly payment.

$$\text{One monthly payment} = 0.7\% \text{ of } \$65{,}000 = \frac{0.7}{100} \times \$65{,}000 = \frac{7}{1{,}000} \times \$65{,}000 = \$455$$

Answer the Question

Step 3: Tandy can expect his monthly payments to be approximately $455.

Now try these problems.

1. Claudette is buying a house that is valued at $35,000. She can make a $5,000 down payment. She needs to borrow the rest. She uses the following rule for a rough approximation of her monthly payments: For a 30-year loan at an 8% interest rate, the monthly payment is about 0.7% of the amount. What is an approximation of Claudette's monthly payment by this rule?

   Answer: Claudette can expect her monthly payment to be

   approximately $__________.

2. Gene can make a $20,000 down payment on a house valued at $70,000. He estimates his monthly payments for a 30-year loan on the remainder to be about 0.7% of the amount he borrows. He

asks his realtor about a 15-year mortgage. His realtor says that if Gene can afford \$100 more each month, he can afford a 15-year mortgage. What is an approximation of Gene's monthly payment for a 15-year mortgage?

A \$140 **B** \$350 **C** \$450 **D** \$490

3. Mort asks his bank representative about getting a mortgage loan. His bank representative tells him that they use 28% of his gross salary to estimate the monthly payment he can afford on a loan. Mort's gross salary, including bonuses and before taxes, is \$24,000. One rule says that the monthly payments on a 30-year loan at an 8% interest rate is about 0.7% of the loan amount. What size loan can Mort get? Use *d* for the loan. Use *p* for his monthly payment. Fill in the blanks to find the answer.

$p \approx$ ______% × (\$24,000 ÷ 12) bank's estimate of monthly payment

$\approx$ \$____________

p $\approx 0.7\% d$ rule's estimate of monthly payment

so $d \approx p \div 0.7\%$

$d \approx$ \$____________ ÷ 0.7% substitute the bank's value for *p*

so $d \approx$ \$____________ the approximate size of the loan Mort can get

4. Jassica's bank charges 1% of any mortgage loan as a fee. Jassica gets a \$60,000 mortgage. Shade the number of squares that represents the fee she must pay if all 100 squares together represent \$10,000.

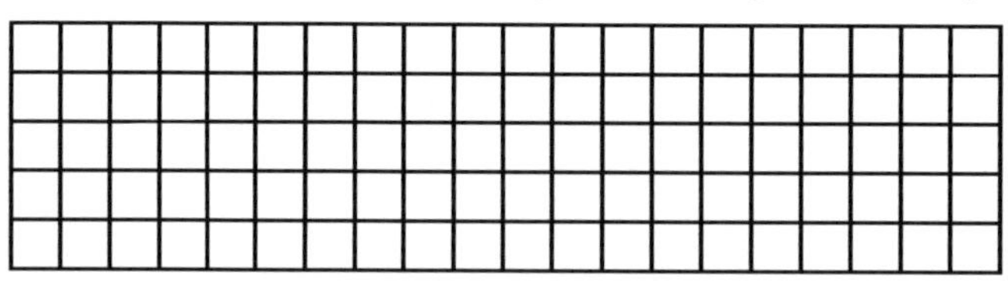

☆ *Challenge Problem*

You may want to talk this one over with a partner.

Eduardo and Malika's house is assessed at a value of \$65,000. Compute the total real-estate taxes they pay, given the tax-rate table.

Taxing Body	**Tax rate**	**Taxes owed**
State	0.02%	
County	0.62%	
Town	0.44%	
School District	1.47%	
Library	0.12%	
TOTAL ANNUAL TAXES		

Lesson 3

College Costs

Example Linden is going to a dental technical college. To cover his expenses, he borrows \$5,000 for three years at 7.5% interest rate per year, compounded annually. Make a table that shows how much he owes at the end of each year.

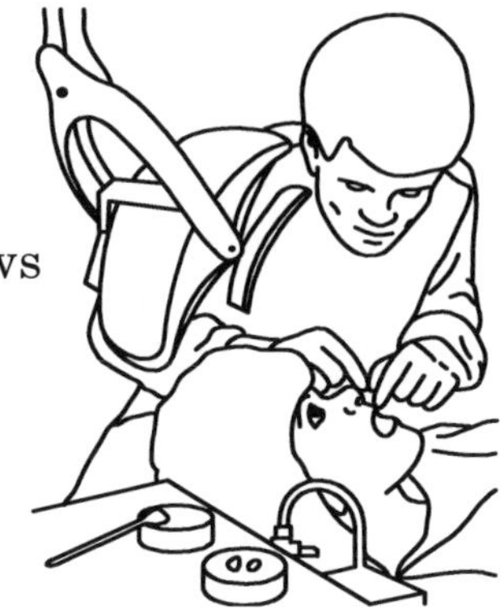

Solve

Step 1: First, define these four column headings for the table.

Year, Balance at start of year, Interest due, Balance at end of year.

Step 2: For year 1, calculate the interest due and the balance at the end of the year.

year 1:

interest due = 7.5% of \$5,000 — yearly interest rate × original loan.

= 0.075 × \$5,000

= \$375

balance at end of year = \$5,375 — balance at start + interest due

Step 3: Repeat the process for years 2 and 3.

year 2:

interest due = 7.5% of \$5,375 — rate × balance at start of year 2

= \$403.13

balance at end of year = \$5,778.13 — balance at start of year 2 + interest

year 3:

interest due = 7.5% of \$5,778.13

= \$433.36

balance at end of year = \$6,211.49 — balance at start of year 3 + interest

Answer the Question

Step 4:

Year	Balance at start of year	Interest due	Balance at end of year
1	\$5,000	\$375	\$5,375
2	\$5,375	\$403.13	\$5,778.13
3	\$5,778.13	\$433.36	\$6,211.49

Now try these problems.

1. Melody is going to a fashion merchandising school. To cover her expenses, she borrows $4,000 for two years at 8.5% yearly interest rate, compounded annually. How much will she owe at the end of the second year?

   A $340 **B** $4,340 **C** $4,680 **D** $4,708.90

2. Marissa loans her nephew $3,000 for a 3-month trip to Europe. She charges him 5% yearly simple interest for the two years that he owes her the money. Draw a pie chart that relates these three amounts: the interest, the original loan and the total amount he owes after two years.

3. Asami wins a scholarship for $6,000 to be used for enrichment courses. The first year she uses $2,500. She invests $2,000 in a bank and earns 2% simple interest annually. She loans the rest to a friend for one year at 5% simple interest. At the end of the first year, how much can she expect to have available to take more courses? Arrange these pieces of the puzzle to create an equation that gives this amount. Fill in the blank puzzle piece with that amount.

   [$2,000] [=] [+] [1.02] [$______]
   [×]
   [1.05] [$1,500] [×]

   Answer: ______________________________

4. Before going to technical school, Rubin works at a computer store. For three years, he puts $1,500 into a savings account at the first of each year. His savings account earns 5% simple interest per year. At the end of each year, he takes out the interest and puts it into a special envelope. Which expression gives the total amount he will have at the end of three years in his account and his envelope?

   A 3(1,500 × 1.05) + 2($1,500 × 1.05) + ($1,500 × 1.05)
   B 3(1,500 × 0.05) + 2($1,500 × 0.05) + ($1,500 × 0.05)
   C $1,500 + 5(1,500 × 0.05) + ($1,500 × 1.05)
   D $4,500 + 6(1,500 × 0.05)

☆ *Challenge Problem*

You may want to talk this one over with a partner.

One report says it will cost about $237,000 (including inflation) to raise a child born in 1999, over the next 17 years. (These costs include food, shelter, and other necessities.) Does this mean that the costs in the first year in high school will be $13,941? Explain your answer.

____

____

Lesson 4 — What Do You Owe?

A Board Game (for Two or More Players)

The goal of this game is to be the first player to reach 25 points or more. First, you take the top flowchart card. Then you apply a number to the flowchart and calculate the answer.

Materials

Flowchart cards (on the next page; players can make up more), and two number cubes

Directions

1. Shuffle flowchart cards, and place them, face down, in the center of the table. Sit with players around the table. One player records player scores.
2. Player 1 tosses the number cubes and adds the two numbers. Player 1 then takes the top flowchart card. This player uses the sum in the flowchart as the value of *INPUT,* and calculates the *OUTPUT.* Other players check the answer. If the answer is correct, the player gets 2 points. If it is incorrect, the first player who calculates the correct value for *OUTPUT* gets 1 point. The flowchart card is discarded to a separate deck.
3. Player 2 then takes a turn. This player can choose to toss the number cubes again, or to use player 1's *OUTPUT* as the next *INPUT* value. Player 2 takes one card from the flowchart card deck and calculates the *OUTPUT.*
4. Players take turns taking a flowchart card, applying a number, and calculating the *OUTPUT.* Replace the flowchart card deck with the shuffled discard deck.
5. The game is over when one player's score is equal to or greater than 25 points.

Before you play the game, try these warm-up problems.

1. Toni tosses the number cubes and gets 9. Her flowchart card says to multiply by 2. If the result is greater than or equal to 40, multiply by $\frac{1}{8}$ for the *OUTPUT.* Otherwise multiply by 2 again, until the result is greater than or equal to 40, and then multiply by $\frac{1}{8}$ for the *OUTPUT.* What is Toni's *OUTPUT?*
2. Andy starts with Toni's *OUTPUT.* His flowchart card says to add 16. If the result is greater than 24, multiply by 40%. In both cases, then divide by 2. What is his *OUTPUT*?

   A 2 **B** 3.5 **C** 5 **D** 10

Flowchart cards for *What Do You Owe?*

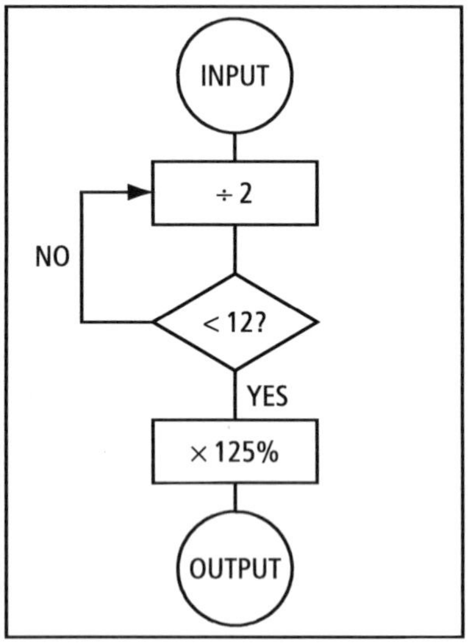

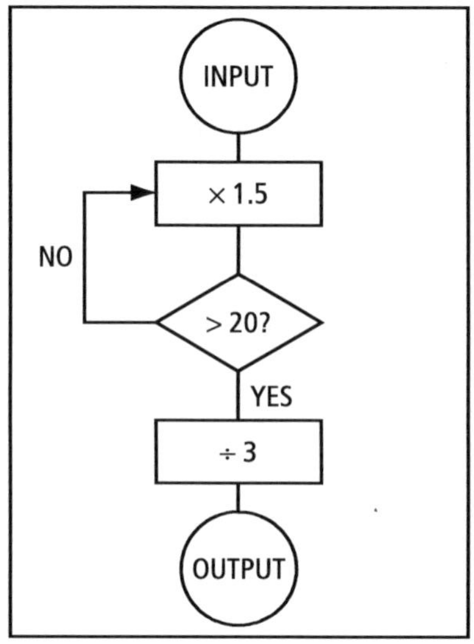

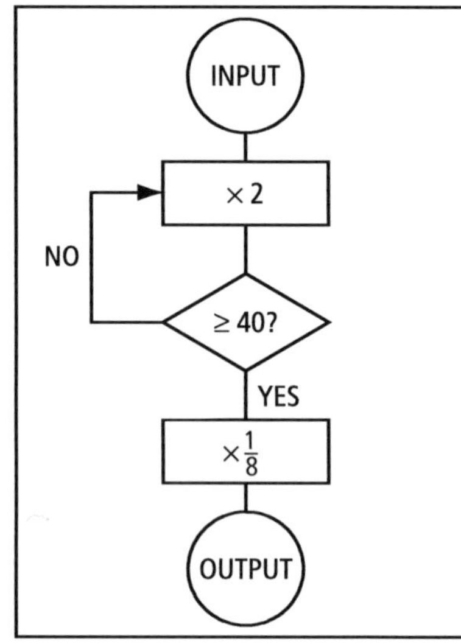

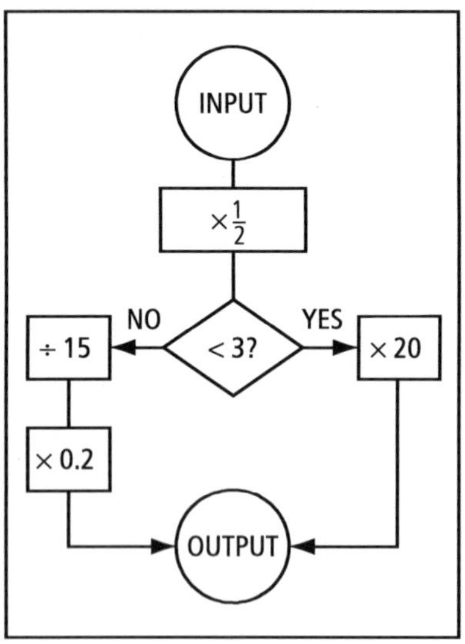

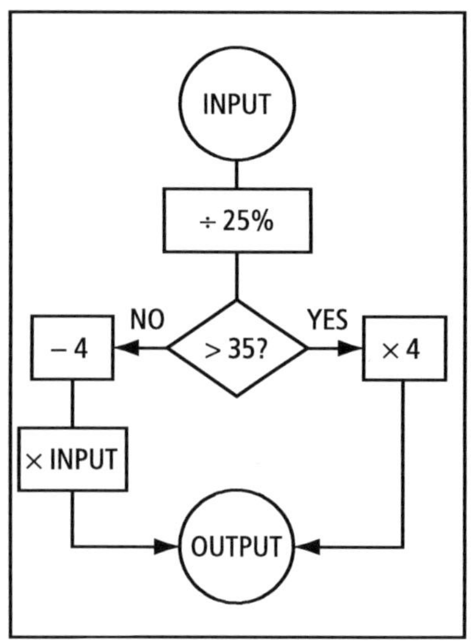

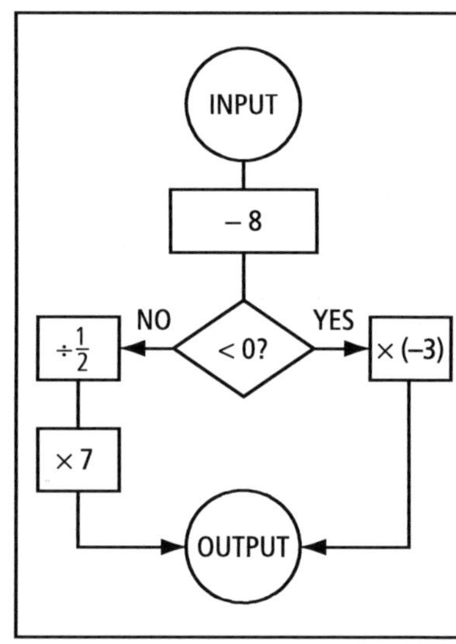

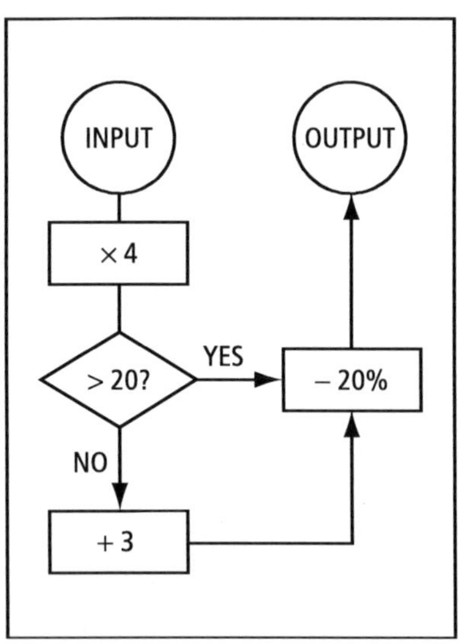

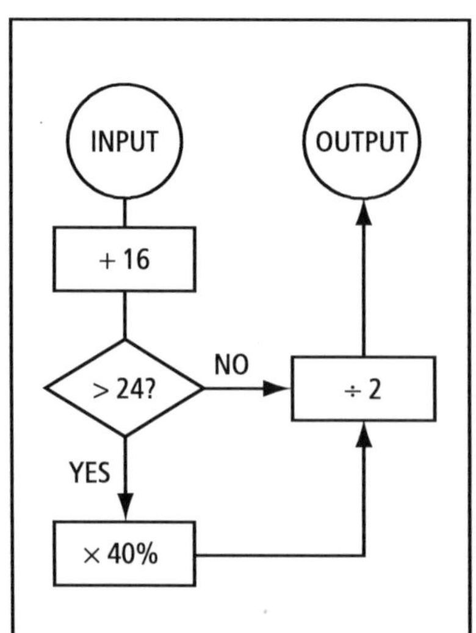

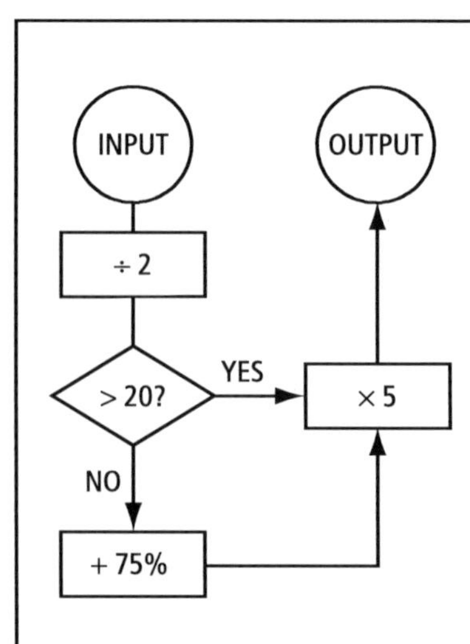

Lesson 5

Buying Insurance

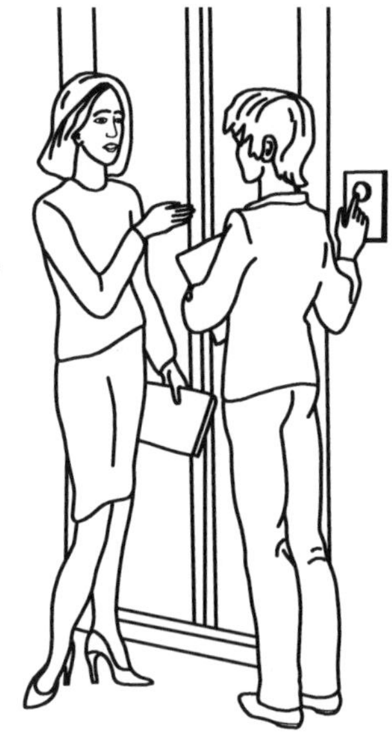

Example Talaitha spends $260 of her monthly income on health insurance. She earns $34,320 per year. Annika spends the same fraction of her monthly income on health insurance. She earns $22,440 per year. How much does Annika spend on health insurance each month?

Solve

Step 1: First, find what fraction of Talaitha's monthly income she spends on health insurance. Calculate her income per month. Then find the fraction.

$\$34,320 \div 12 = \$2,860$

$\frac{\$260}{\$2,860} = \frac{1}{11}$ health insurance payment as a fraction of monthly income

Step 2: Now, find the same fraction of Annika's income.

$\frac{1}{11}$ of $\$22,440 = \$2,040$ This is Annika's *annual* health insurance.

$\$2,040 \div 12 = \$170$

Answer the Question

Step 3: Annika spends $170 on health insurance each month.

Now try these problems.

1. Reed spends $133 of his monthly income on car insurance. He earns $14,364 per year. Luther spends the same fraction of his monthly income on health insurance. He earns $17,280 per year.
   a. What fraction of his monthly income does Luther spend on health insurance?
   b. How much does Luther spend on health insurance each month?

   Answer: a. Luther spends ______ of his monthly income on health insurance.

   b. Luther spends $_______ per month on health insurance.

2. For $500, Cat can buy insurance for her new house. But this insurance does not include insurance against water damage from broken pipes. For $1,600 she can buy insurance that does cover water damage. Suppose she buys the policy that does not include

insurance against water damage. Now suppose she goes three years without water damage, and puts the difference in a savings account. How much does Cat save?

A $1,500 **B** $3,300 **C** $4,800 **D** $6,300

3. Jerilee buys a basic health insurance policy. The company offers a rider costing $24 per month that will pay for office visits. Her doctor's office tells her that the average cost for a visit is $80. Usually Jerilee visits the doctor twice a year. Answer the questions, and state your conclusion.

Answer: a. For the office visit rider, Jerilee pays $______ this year.

b. For two office visits (without the rider) she pays $______.

c. Jerilee would have saved money by __________ (buying/ not buying) the office visit rider.

4. Davis buys a new bike for $285. It has a 5-year guarantee. Twelve months later it is stolen from the garage. Davis makes an insurance claim based on a fraction of the bike's original price. This fraction is equal to the time left on the guarantee divided by the total years of the guarantee. Make a table to show the value of the bike after 1, 2, 3, 4, and 5 years. Circle the value of the bike that Davis can expect the insurance company to pay.

Years after purchase	Time left on guarantee / total years	Value
1		
2		
3		
4		
5		

☆ *Challenge Problem*

You may want to talk this one over with a partner.

You have basic car insurance that does not include towing. Last year, your truck broke down twice, and you paid $50 and $110 to get towed. For an extra $18 per year you can add towing coverage to your policy. The insurance company will then pay up to $75 per towing. Would you pay the extra $18 to add towing coverage to your policy? Explain your answer.

Unit 2 Review

Review What You Learned

In this unit you have used mathematics to solve many problems. You have used mental math and estimation, practiced basic operations, solved equations, and used statistics, ratios, and proportions.

These two pages give you a chance to review the mathematics you used and check your skills.

✔ Check Your Skills

1. Audrey finally decides which car to buy. The salesperson offers her two 0%-interest payment plans. These are the details.

   i. With no down payment, she pays for the car in twenty-four equal monthly installments.
   ii. With a down payment of \$2,500, she pays \$50 less per month over twenty months.

   Write two equations to find the price of the car. What does Audrey pay per month with no down payment? What is the price of the car?

   Answer: ______________________ ______________________

   Audrey pays \$__________ per month with no down payment.

   The price of the car is \$__________.

   If you need to review, return to lesson 1 (page 15).

2. Moneya is buying a house that is valued at \$45,000. She can make a \$5,000 down payment. She needs to borrow the rest from a loan company. For a 30-year loan at an 8% interest rate, her monthly payments will be about 0.7% of the amount she owes. Which expression gives an approximation of Moneya's monthly payments?

   A $0.7 \times \$40,000$ **C** $0.007 \times \$40,000$

   B $0.07 \times \$40,000$ **D** $0.0007 \times \$40,000$

   If you need to review, return to lesson 2 (page 17).

3. Brody is going to technical school. To cover his expenses, he borrows \$6,000 for two years at 6.5% yearly interest rate, compounded annually. He makes a table that shows how much

he will owe at the end of each year. Are his figures correct? Make any necessary changes.

Year	Balance at start of year	Interest incurred	Balance at end of year
1	$6,000	$390	$6,390
2	$6,390	$830.70	$7,220.70

If you need to review, return to lesson 3 (page 19).

4. James spends $142.92 of his monthly income on car insurance. He earns $14,292 per year. Hal spends the same fraction of his monthly income on car insurance. He earns $16,500 per year.

a. Draw a diagram to show what fraction of his monthly income Hal spends on car insurance.

b. How much does Hal spend on car insurance each month?

Diagram a.

Answer: b. Hal spends $________ on car insurance each month.

If you need to review, return to lesson 5 (page 23).

5. Kay buys a basic health-insurance policy. The company offers a rider for office visits for $15 per month. The average cost for an office visit is $75. She usually visits the doctor three times a year.

Kay would save money by ______________ (buying/not buying) the office visit rider.

If you need to review, return to lesson 5 (page 23).

Write Your Own Problem ✍

Choose a problem you liked from this unit. Write a similar problem using a situation and related facts from your own life. With a partner, share and solve these problems together. Discuss the mathematics and compare the steps you used. If you need to, rewrite or correct the problems. Write your edited problem and the answer here.

____

____

____

____

____

Unit 3 Unplanned Expenses

Preview

How You Will Use This Unit

Unplanned expenses may involve many different things. Unexpected changes in rent, bills, medical or dental care, and unexpected repairs are some examples. Unanticipated gifts and weekend trips are other examples. As you handle unplanned expenses, you will often use math. The math skills you use include mental math and estimation, basic operations and equations, statistics, and ratios and proportions.

What You Will Do in This Unit

In this unit, math steps demonstrate how to solve problems. These steps can help you answer questions such as these:

Last winter, your electricity bill was $85 per month. In the summer, this amount doubled. You now budget the maximum amount each month. How much do you budget for November?

You go online to look for the prescription drug that your doctor recommends. The drug costs $56. Priority mail is free for orders over $49, and $8.99 for second-day service. How much less do you pay if you order by priority mail?

You get another heel dent in your sailboard. Last week, you got a heel dent repaired for $300. You think that this one is only half as bad as the last one. What can you expect to pay this time?

You buy a cake, a card, and a CD for a friend's birthday. The cake costs one third of the money in your wallet. The CD costs as much as the cake. The card costs one quarter the cost of the cake. What fraction of the money that you start with do you have after buying everything?

What You Can Learn from This Unit

When you complete this unit, you will have used mathematics to work problems related to unplanned expenses. These problems are similar to those that may actually occur in your daily life.

Lesson 1 → *Telephone & Other Bills*

Example Last summer, Charlene's average electricity bill was about $99 per month. During the winter, her average electricity bill was about $150 per month. She had to borrow money to pay for this unexpected increase. Charlene now budgets each month the higher average monthly amount. She also budgets for the recent 5% increase in electricity rates. How much does she budget for electricity for next July?

Solve

Step 1: Underline the words that tell you the higher average amount that Charlene paid per month for electricity last year.

<u>... her average electricity bill was about $150 per month.</u>

Step 2: Underline the sentence that tells you about any changes in electricity rates.

<u>She also budgets for the recent 5% increase in electricity rates.</u>

Step 3: Calculate the monthly amount that Charlene budgets for this year, including the increase.

105% of \$150 = 1.05 × \$150
= \$157.50

Step 4: Draw a conclusion about what Charlene will budget for electricity in July.

\$157.50 She budgets the same amount each month.

Answer the Question

Step 5: Charlene budgets $157.50 for electricity in July.

✏ Now try these problems.

1. Last year, Alfonso's average telephone bill was about $65 per month. Since he moved away to tech school, his average telephone bill increased to about $110 per month. Last month, he had to borrow money to cover this unexpected increase. He now budgets the higher amount per month to cover his telephone bill. What is his annual budget for telephone use now?

   Answer: Alfonso's annual budget for telephone use is $__________.

2. Libby is surprised to see a bank charge of $17.20 on her bank statement this month. The bank says that she did not maintain a minimum daily balance of $2,000. This means that she must pay a $3 service fee plus 10 cents for each transaction. There is also a fee of $12 for a lost check on which she stopped payment.

   a. How many transactions did she have this month? Circle that amount.

      10 22 30 52 172

   b. How much did it cost her to have her daily balance fall below $2,000? Draw a rectangle around that amount.

      $3 $3.10 $5.20 $12 $17.20

3. This month, the air conditioning costs Melvyn an extra $75 to run. Last month, his telephone bill cost him an extra $120. The month before that, he paid $45 to replace a lost contact lens. Imagine you are Melvyn. On average, how much would you put aside per month for unexpected expenses like these? Explain your answer.

   A $45 **B** $75 **C** $80 **D** $120

   Explanation: ______________________________

4. Each September, Berta renews her rental contract. This year her rent goes up by 10% from the previous year's $545 per month. Each box represents $54.50. Shade the number of boxes that show her new rent.

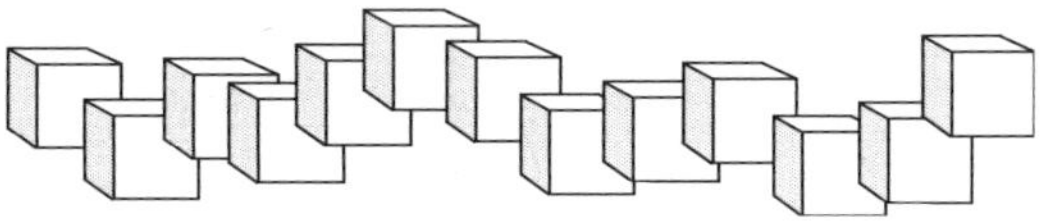

☆ *Challenge Problem*

You may want to talk this one over with a partner.

You and your friend each have a number. These two numbers add to 5, 10, or 21. You have 7. You ask if your friend knows your number. Your friend replies, No. What number does your friend have? How do you know that number?

____

____

____

Lesson 2 → Doctors, Dentists, & Veterinarians

Example Pierce goes online to look for the diabetes testing kit that his doctor recommends. He finds a website that lists the kit. It costs $74 and weighs 2 lbs. He consults the table of shipping rates on that website. How much *less* does he pay for 5–7 day priority mail than for next-day service?

Shipping Rates	
5–7 day priority mail/UPS Ground	Free (for orders over $49)
5–7 day priority mail/UPS Ground	$3.99 (for orders under $49)
Second-day service	$8.99 + $1.50 per lb over 1 lb
Next-day service	$14.99 + $3.00 per lb over 1 lb

Solve

Step 1: Underline the line in the table that tells the rate for next-day service. Then find the exact rate that Pierce would pay for this service.

Next-day service $14.99 + $3.00 per lb over 1 lb

Pierce would pay $17.99 His order weighs 2 lbs.

Step 2: Pierce's order is over $49. Find the line for 5–7 day priority mail when an order is that size.

5–7 day priority mail/UPS Ground free (for orders over $49)

Step 3: Write an expression for the difference in rates.
$17.99 – $0 = $17.99

Answer the Question

Step 4: Pierce pays $17.99 less for 5–7 day priority mail than for next-day service.

Now try these problems.

Use the shipping rates table above.

1. Chase goes online to look for the prescription drug that her veterinarian recommends. The website lists the drug for $25, and the package weighs less than 1 lb. She consults the table of shipping rates. How much *more* does she pay for next-day service than for second-day service?

   Answer: Chase pays $__________ more for next-day service than for second-day service.

2. Carolee has to go out of town unexpectedly. She asks her veterinarian if he knows of a cat-sitting service. The veterinarian says his assistant will make a pet visit twice a day for $10 per day. Carolee pays him $50. How many visits does he make?

   A 5 **B** 10 **C** 15 **D** 20

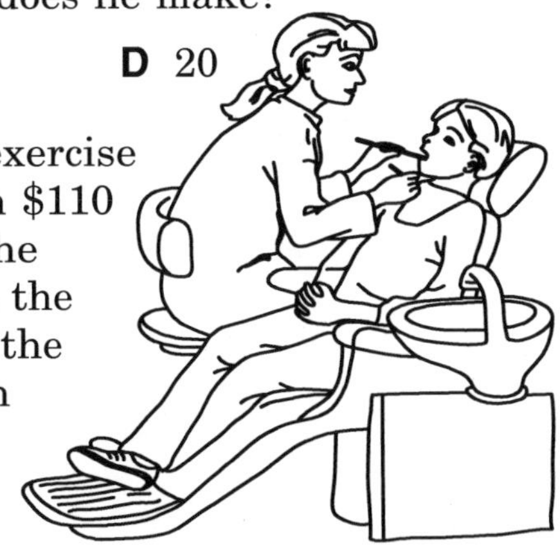

3. Philo cracks his tooth in an accident on the exercise equipment at school. The dentist charges him $110 for X-rays and $650 for a cap on the tooth. The charge is $35 for a final check-up. Philo pays the total bill in equal monthly installments over the next six months. How much does Philo pay in each installment? Complete the following expression to show this amount.

   ($110 + $________ + $________) ÷ ____ = $________

4. Shelley goes to the hospital emergency room for a stomach problem. The visit costs her $135. Her insurance deductible is $2,000. So far, she has paid $1,365 of this deductible. Circle the drawing that shows the percent of her deductible left after she pays the $135 bill.

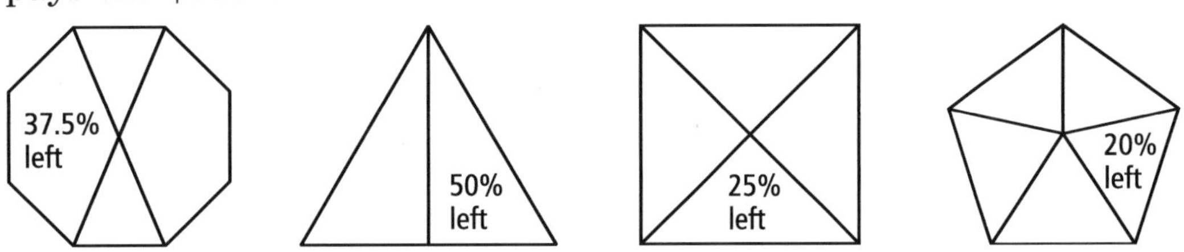

☆ *Challenge Problem*

You may want to talk this one over with a partner.

Look at the figure below that shows the first 5 rows of a number triangle. Find the mean of the first and last term for each row. Write this mean to the right of each row. Predict the mean of the first and last term for row 100.

Answer: The mean of the first and last term for row 100 is __________.
Explain your reasoning.

____

____

____

____

1 ________
3 5 ________
7 9 11 ________
13 15 17 19 ________
21 23 25 27 29 ________

Lesson 3 — Careful Spending

A Board Game (for Two or More Players)

The goal of this game is to cross out two numbers in each row and each column, until each row and column is left with four numbers that add to 10. First, you check numbers in a row and in an intersecting column. Then you cross out the number at the intersection, as long as this moves you one step closer to the goal of the game.

Materials

Game board (on the next page) and pencils

Directions

1. Give each player a pencil. Sit with players around the game board.
2. Following these two rules, player 1 crosses out one number on the board:
   (i) Four or more numbers must remain in the row and in the column after the number is crossed out.
   (ii) Check the sum of the uncrossed-out numbers in each row and each column. It must be possible for the four numbers remaining at the end of the game to add to 10.
3. Players take turns testing and checking numbers on the board. Each player crosses out one number to approach the goal of the game.
4. The game is over when the goal of the game is achieved: (a) the four numbers remaining in each row add to 10 and (b) the four numbers remaining in each column add to 10.

Before you play the game, try these warm-up problems.

1. Paulette crosses out the 7 in row 4, column 2. Which numbers can be crossed out in row 4 or column 2 to satisfy the rules of *Careful Spending*?

   Answer: ______ in row 4, and ______ in column 2.

2. Ralph follows Paulette's move. He crosses out the 1 in row 1, column 2. Sara follows Ralph. What number can she cross out to satisfy the rules of *Careful Spending*?

   A 5 in row 1, column 5
   B 4 in row 1, column 6
   C 2 in row 1 column 1
   D 2 in row 1, column 3

Game Board for *Careful Spending*

2	1	2	2	5	4
5	1	1	6	2	3
3	3	5	1	3	1
1	7	6	1	1	2
2	5	1	4	1	3
4	1	2	4	2	3

Lesson 4 — Unplanned Repairs

Example Alton blows out his windsurfer sail during the beach festival. *On-the-Spot Repairs* advertises that they can repair sails inexpensively and overnight. Alton's friend Court says that the cost of sail repair is approximately proportional to the size of the sail. Court's sail is $\frac{2}{3}$ the size of Alton's sail. Court paid $300 to have his sail repaired. What can Alton expect the approximate cost of his repair to be?

Solve

Step 1: Use x for the size of Alton's sail. Write a ratio that compares the size of Alton's sail to Court's sail.

Alton's ÷ Court's $x \div \frac{2}{3}x$ which simplifies to $1/(\frac{2}{3})$ or $\frac{3}{2}$

Step 2: Underline the words that tell you how size and cost of repair are related.

<u>... the cost of sail repair is approximately proportional to the size of the sail.</u>

Step 3: Write a ratio to compare Alton's repairs (y) to Court's repairs.

Alton's ÷ Court's $\frac{y}{\$300}$

Step 4: Write a proportion by setting the two ratios equal.

$\frac{\text{A's size}}{\text{C's size}} = \frac{\text{A's cost}}{\text{C's cost}}$ so $\frac{3}{2} = \frac{y}{\$300}$

Step 5: Solve the proportion for y, the cost of repairing Alton's sail.

$\frac{3}{2} = \frac{y}{\$300}$ Cross multiply.

$2y = \$300 \times 3$

$2y = \$900$ Divide each side by 2.

$y = \$450$

Answer the Question

Step 6: Alton can expect the cost of his repair to be about $450.

Now try these problems.

1. A storm damages Jody's computer. She pays $180 to have it repaired. She also buys a surge protector. The cost of the surge protector is $\frac{1}{5}$ of the total bill (repairs and protector). Complete this equation you could use to solve for the cost of the surge protector, x.

   ____________ / ____________ $= \frac{1}{5}$

2. The garage tells Ryanna that they can replace the bug shield catch on her car for $130. The bug shield is an optional accessory. Ryanna has the bug shield removed at no cost. She sells the bug shield to her brother for $20. How much does she save? Explain your answer.

   Answer: ____

3. Tyson pays $85 to get his truck towed to the nearest repair shop. The shop keeps the truck for 2 days, and charges him $150 to make the repair. A rental car would cost about $45 per day. Tyson has insurance to cover the cost of a rental car, but not to cover towing. How much does the problem with his truck cost him per day?

   A $117.50 **B** $162.50 **C** $235 **D** $280

4. On the coldest day of the year, Verity's furnace breaks. The repair costs her $520. Verity's emergency fund has $350 in it. She adds $25 to the fund each month. The repair shop allows her to pay $100 right away, and $50 per month without interest. Complete the table and fill in the blanks.

Month		0	1	2	3	4	5	6	7	8	9	10	11
Balance of repair bill	$520	$420											
Emergency fund	$350												

   a. It takes Verity ______ months to pay off the bill.

   b. Verity has $________ left in her emergency fund when the repair bill is paid off.

☆ *Challenge Problem*

You may want to talk this one over with a partner.

You drive an old car. You have an unexpected fender-bender accident. A local body shop gives you a quote of $250 to replace the fender. A friend tells you to take the car to the local vocational tech school. They pound out the fender and repaint it. They charge $20. What sort of donation do you make to the school?

____

____

Lesson 5 → *Gifts, Flowers, & Unexpected Trips*

Example Lena promises to pick up a cake for her best friend's birthday party. The cake costs one third of the money that she has in her wallet. She buys a birthday card that costs one fourth of the cost of the cake. She also buys a gift of a CD that costs as much as the cake. What fraction of the money that Lena starts out with does she have after buying everything?

Solve

Step 1: Use x for the amount of money that Lena has in her wallet when she starts out. Write an expression for the cost of the cake.

$\frac{1}{3}x$ one third of the money

Step 2: Next, write expressions for the cost of the card and the cost of the CD.

Card: $\frac{1}{4}(\frac{1}{3}x)$ one fourth of the cost of the cake which simplifies to $\frac{1}{12}x$

CD: $\frac{1}{3}x$ equal to the cost of the cake

Step 3: Now, write an expression for the money that Lena has left over afterwards.

$x - (\frac{1}{3}x + \frac{1}{12}x + \frac{1}{3}x)$ Find the common denominator for the fractions and add.

$x - (\frac{4}{12} + \frac{1}{12} + \frac{4}{12})x$

$x - \frac{9}{12}x$

$x - \frac{3}{4}x$

$\frac{1}{4}x$

Answer the Question

Step 4: Lena has $\frac{1}{4}$ of the money she starts out with after buying everything.

Now try these problems.

1. Faulkner buys paper, tape, and glue for his project. The paper costs one half of the money in his wallet. The tape and the glue each costs one third of the cost of the paper. Faulkner also picks up cookies for half the cost of the tape.

a. What fraction of the money that Faulkner starts out with does he have afterwards?

b. This line represents the money that Faulkner starts out with. Mark distances on the line, one after the other, to show the fraction that each purchase represents.

Answer: a. Faulkner has ______ of the money that he starts out with.

b. |____|

2. Marisela decides to buy gift items when she sees them on sale. This way, she has gifts ready for birthdays, and she also saves money. On Saturday, she buys items marked with these sale prices: a book for \$5.98, a game for \$9.99, and a candle for \$4.59. These prices are half of the normal price. How much does she save?

   A \$6.85 **B** \$10.28 **C** \$20.56 **D** \$41.12

3. On Tuesday morning, Javier's boss asks Javier to go to a meeting in town. The company pays for Javier's mileage and lunch. Javier stays in town and eats supper with friends. He spends \$12.75. He usually spends \$5 for lunch out and \$4 to prepare supper. How much more or less than usual did he spend for this special day? Explain your answer.

   Answer: He spent \$________ ________ (more/less).

4. Ane reviews last year's pattern of unplanned expenses. She spent about \$18 per month on flowers, except for two months when she spent \$72.50 and \$59.50. She expects this pattern to continue. What is the average (mean) amount that she should budget per month for flowers next year? Complete this expression to show the amount.

   Answer:

   (\$18 × ________ + ________ + ________) ÷ ________ = \$________

☆ *Challenge Problem*

You may want to talk this one over with a partner.

You go online and find that you can send a gift plant overnight, direct from the growers. One choice is a Chinese Elm Bonsai tree, delivered, for only \$43. The normal price for this kind of plant is \$64.50. What fraction of the normal price is the advertised price? Explain why, in your opinion, you often pay less for things over the Internet.

____

____

____

Unit 3 Review

Review What You Learned

In this unit you have used mathematics to solve many problems. You have used mental math and estimation, practiced basic operations, solved equations, and used statistics, ratios, and proportions.

These two pages give you a chance to review the mathematics you used and check your skills.

✔ Check Your Skills

1. Last winter, Caylor's average electricity bill was about $85 per month. During the summer, his average electricity bill was about $175 per month. Caylor now budgets the higher monthly amount each month. He also allows for the recent 4% increase in electricity rates. How much does he budget for electricity this November?

   Answer: $__________.

   If you need to review, return to lesson 1 (page 28).

2. Nancy goes online to look for medication that her doctor recommends. She finds a website where it costs $79, and the package weighs less than 1 lb. She consults the table of shipping rates for the website. How much *less* does she pay for 5–7 day priority mail than for second-day service? Fill in the blanks with the data. Answer the question.

Shipping Rates	
5–7 day priority mail/UPS Ground	Free (for orders over $49)
5–7 day priority mail/UPS Ground	$3.99 (for orders under $49)
Second-day service	$8.99 + $1.50 per lb over 1 lb
Next-day service	$14.99 + $3.00 per lb over 1 lb

   Answer: Nancy would pay $__________ for 5–7 day priority mail.

   OR she would pay $__________ for second-day service.

   Nancy would pay $__________ less for 5–7 day priority mail.

   If you need to review, return to lesson 2 (page 30).

3. Linden gets another heel dent in his sailboard. Last week, *On-Time Repairs* repaired a heel dent for him for $300. Linden thinks that this new dent is only half as bad as last week's dent. He expects

the cost of the new repair to be roughly proportional to the previous one. What can he expect to pay to get this new heel dent repaired?

A \$75 **B** \$150 **C** \$300 **D** \$450

If you need to review, return to lesson 4 (page 34).

4. Delfino buys sandwiches, chips, and drinks for the team after the ball game. The sandwiches cost one half of the money he has in his wallet. The chips and the drinks each cost one quarter of the cost of the sandwiches. What fraction of the money that Delfino starts out with, does he have after buying everything? Draw and label a diagram to show this amount as part of the whole.

If you need to review, return to lesson 5 (page 36).

Write Your Own Problem

Choose a problem you liked from this unit. Write a similar problem using a situation and related facts from your own life. With a partner, share and solve these problems together. Discuss the mathematics and compare the steps you used. If you need to, rewrite or correct the problems. Write your edited problem and the answer here.

Unit 4 Credit Card Budgeting

Preview

How You Will Use This Unit

Credit card budgeting involves many different things. Understanding how a credit card works and what the fine print really says, are two examples. Paying bills by credit card and using credit card services are others. As you compare options and make choices, you will often use math. The math skills you use include mental math and estimation, basic operations and equations, statistics, and ratios and proportions.

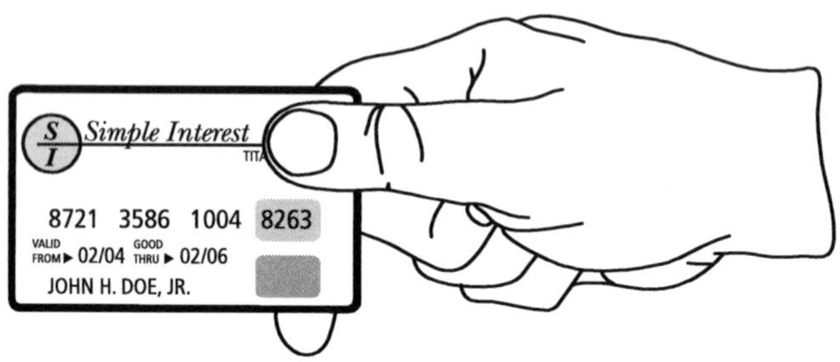

What You Will Do in This Unit

In this unit, math steps demonstrate how to solve problems. These steps can help you answer questions such as these:

You buy a keyboard for $455 and charge it on your credit card. The card company charges 1.9% monthly interest on unpaid balances. After a first payment of $25, what do you owe?

Your credit card offers frequent-flyer miles for an annual fee of $50. Your goal is a 20,000 mile ticket. You charge about $2,500 each year. How much will your ticket cost in annual fees alone?

The closing date for one credit card is the 25th of the month. The closing date for another card is the 15th of the month. You make a purchase on the 7th of the month. Which card allows you use of the money longer?

On May 16th you mail $20 of your $156.50 balance to the credit card company. The closing date is the 16th. Payments not received by that date will cause a late fee. Late fees are $20 on balances up to $150 and $35 on balances over $150. How much do you now owe?

What You Can Learn from This Unit

When you complete this unit, you will have used mathematics to work problems related to credit card budgeting. These problems are similar to those that may actually occur in your daily life.

Lesson 1 — How Credit Cards Work

Example Jose buys a boom box for $350 and charges it on his credit card. At the end of the first month, he makes a $50 payment. The credit card company charges 1.5% monthly interest on unpaid balances. After Jose's $50 payment is deducted, interest is computed on the unpaid balance.

a. How much interest does the credit card company charge Jose?

b. This interest is added to the unpaid balance. How much does Jose now owe?

State Bank Credit Card Services			ACCOUNT NAME José Lopez	
PREVIOUS BALANCE	PAYMENTS RECEIVED	NEW PURCHASES	INTEREST	UNPAID BALANCE
$100	$100	$350	$0	$350

DATE OF PURCHASE	ITEM	PRICE
10/1/03	Boom box	$350

Solve

Step 1: First, find the amount that Jose has left to pay after his first month's payment.

$350 – $50 = $300

Step 2: Next, underline the sentence that tells the interest rate that the credit card company charges on unpaid balances.

<u>The credit card company charges 1.5% monthly interest on unpaid balances.</u>

Step 3: Now, find the interest on Jose's unpaid balance.

1.5% of $300
0.015 × $300
$4.50

Step 4: To find how much money Jose owes at the end of the first month, add.

$300 + $4.50 = $304.50   Add the interest to the unpaid balance.

Answer the Question

Step 5: **a.** The credit card company charges Jose $4.50 interest.
b. He now owes $304.50.

Now try these problems.

1. Kelly buys a new outfit for $212, and charges it on her credit card. At the end of the first month, she makes the minimum payment of

$10. The credit card company charges 2.1% monthly interest on unpaid balances. After Kelly's $10 payment is deducted, interest is computed on the unpaid balance. How much does Kelly owe?

Answer: Kelly now owes $____________.

2. Mae buys an entertainment center on sale for $325, and charges it on her credit card. The credit card company charges 1.8% monthly interest on unpaid balances. Mae pays $50 per month to the credit card company.

   a. Make a table to show how many months it takes for Mae to pay off her purchase.

Month	1	2	3	4	5	6	7	8	9
Payment	$50								
Amount owed									

   b. How much does she pay for the entertainment center?

Answer: b. Mae pays $____________ for the entertainment center.

3. Duke's bank charges 1.5% monthly interest for a bank credit card. Duke uses his credit card frequently and makes monthly payments every month. But, for a whole year, his unpaid balance stays above $1000. Which estimates the interest that he pays on his unpaid balance over one year?

   A 0.015 × 12 × $1,000 = $180
   B 0.015 × $1,000 = $15
   C 0.15 × 12 × $1,000 = $1,800
   D 0.15 × $1,000 = $150

4. The Ellis family has four card holders on one credit card account. The company charges an annual fee of $10 per credit card holder. The company also charges 1.4% monthly interest on unpaid balances. The family's unpaid balance stays at about $1,500, from month to month. About how much does the Ellis family pay for their credit over one year? Show your calculations.

Answer: ____

☆ *Challenge Problem*

You may want to talk this one over with a partner.

One credit card company charges 12.5% annual interest on unpaid balances. A second credit card company charges 8% annual interest on unpaid balances. The second credit card company charges an annual fee of $35. (The first company charges no fee.) How low does an unpaid balance have to go for the second company's low interest advantage to disappear? Explain your answer.

♠ A Card Game (for Two or More Players)

The goal of this game is to make sets of four equivalent cards. First, a player opens a set with a percent (%) card. Then players build on to that set with equivalent forms of that percent.

Materials

Card deck (on the next page)

Directions

1. Deal six cards to each player. Place the remaining deck of cards face down.
2. Player 1 starts by placing a percent card (a card with a % on it) from their hand face up on the table. Then player 1 draws the top card from the deck to replace the one put down. If player 1 does not have a percent card, the turn passes to the next player, until a player can open a set.
3. Players take turns playing one card per turn (and drawing a replacement card). A player may play a card on any open set or start another set (with another percent card). To play a card on an open set, that card must show an equivalent form of the opening percent card. Any number of sets can be open at once.
4. The player who plays the fourth card on an open set wins the set, and scores 5 points. A wild card may be used in place of any card except a percent (%) card. A player may later replace a wild card on the table with an equivalent card in the set.
5. When the card deck is gone, play continues as long as a player can play a card. The player who goes out first scores 10 points. The winner is the player with the greatest number of points.

✏ Before you play the game, try these warm-up problems.

1. One set is open with the cards 25%, $\frac{25}{100}$, and 0.25. You have the following cards: $\frac{30}{100}$, $\frac{17}{20}$, 0.2, $\frac{1}{4}$, 0.8, and 30%. Which card would you put down and why?

   Answer: ______________________________

2. Zion has a card with $\frac{5}{8}$ on it. On what percent card can he play this card?

   A 20% **B** 30% **C** 62.5% **D** 80%

Card Deck for *Percents*

The card deck is made up of sets of four cards. Each set contains equivalent forms of a percent. (For example, 20%, $\frac{20}{100}$, $\frac{1}{5}$, 0.2, as shown in the card set below.)

Make 13 (or more) sets of four cards for different percent values. Make 2 wild cards. For example, make cards for the following sets of equivalent forms:

1.5%	$\frac{1.5}{100}$	$\frac{3}{200}$	0.015
5%	$\frac{5}{100}$	$\frac{1}{20}$	0.05
10%	$\frac{10}{100}$	$\frac{1}{10}$	0.1
12.5%	$\frac{12.5}{100}$	$\frac{1}{8}$	0.125
20%	$\frac{20}{100}$	$\frac{1}{5}$	0.2
25%	$\frac{25}{100}$	$\frac{1}{4}$	0.25
30%	$\frac{30}{100}$	$\frac{3}{10}$	0.3
45%	$\frac{45}{100}$	$\frac{9}{20}$	0.45
50%	$\frac{50}{100}$	$\frac{1}{2}$	0.5
62.5%	$\frac{62.5}{100}$	$\frac{5}{8}$	0.625
75%	$\frac{75}{100}$	$\frac{3}{4}$	0.75
85%	$\frac{85}{100}$	$\frac{17}{20}$	0.85
90%	$\frac{90}{100}$	$\frac{9}{10}$	0.9

20%	$\frac{20}{100}$	$\frac{1}{5}$	**0.2**

Make two wild cards:

Lesson 3 — *Paying by Credit Card*

Example Denna charges her new bicycle on her credit card. The bike is on sale for $175. (The regular price is $256.) She makes no more purchases with her credit card. At the end of each month she pays $40. The credit card company charges her 1.5% monthly interest on the unpaid balance. Does she save money by buying the bike on sale with her credit card?

Solve

Step 1: Underline the sentences that tell you what Denna pays for the bike and what she pays at the end of each month.

The bike is on sale for $175.
At the end of each month she pays $40.

Step 2: Make a table to show her monthly unpaid balance until the balance is cleared.

End of Month	Denna pays	Remaining Unpaid Balance
1	$40	($175 – $40) × 1.015 = $137.03
2	$40	($137.03 – $40) × 1.015 = $98.49
3	$40	($98.49 – $40) × 1.015 = $59.37
4	$40	($59.37 – $40) × 1.015 = $19.66
5	$19.66	$0

Step 3: Find the total Denna pays for the bike on her credit card. Compare this to the regular price of the bike.

$40 × 4 + $19.66 = $179.66   This is less than the regular price of $256.

Answer the Question

Step 4: Yes, Denna saves money by buying the bike on sale with her credit card.

✏ Now try these problems.

1. Louie charges his new saxophone on his credit card. The saxophone is on sale for $315. (The regular price is $499.) He makes no more purchases with his credit card. At the end of each month he pays $75. The credit card company charges him 2.0% monthly interest on his unpaid balance.

a. Does he save money by buying the saxophone on sale with his credit card?

b. What is the difference between what he pays and the regular price?

Answer: a. ______ (Yes/No). Louie ____________ (saves/does not save) money by buying the saxophone with his credit card.

b. The difference is $____________.

2. *All*Insured* charges Alton $1,596 per year, or $140 per month for car insurance. Alton decides to pay monthly, by credit card. He makes no more purchases with his credit card. At the end of each month he pays $50. The credit card company charges him 1.5% monthly interest on his unpaid balance.

a. Complete the table to show Alton's unpaid balance at the end of each month.

End of month	1	2	3	4	5	6
Unpaid balance						
End of month	7	8	9	10	11	12
Unpaid balance						

b. Alton continues to need car insurance. What is happening to his credit card debt?

Answer: b. ____

3. The closing date for Vandana's *Diamond* credit card is the 10th of each month. The closing date for her *MyCharge* credit card is the 25th of each month. Each card offers a grace period of 30 days. She makes a purchase on the 5th of the month. Which card allows her money to stay in the bank longer? Why?

A *Diamond;* payment due in 5 days
B *Diamond;* payment due in 35 days
C *MyCharge;* payment due in 20 days
D *MyCharge;* payment due in 50 days

☆ *Challenge Problem*

You may want to talk this one over with a partner.

You plan to go to a conference in just over three months from now. When you pay 3 months in advance, the cost is $285. But you will have to pay by credit card. Your credit card charges 2.1% monthly interest. This month you would make a $10 payment. Next month you would pay the balance in full. When you pay 2 months in advance, the cost is $325. By then you would be able to pay in full. Which method of payment would you use? Why?

____

Lesson 4 — Credit Card Services

Example Marius gets a credit card that offers frequent-flyer miles. The annual fee is $50. He will earn one mile for every dollar he charges on his credit card. He reads that it takes about 25,000 miles to earn a free round-trip ticket anywhere in the United States. On average, Marius spends $5,000 in credit card charges per year. How much will it cost him in annual fees alone to earn a free round-trip ticket in the United States?

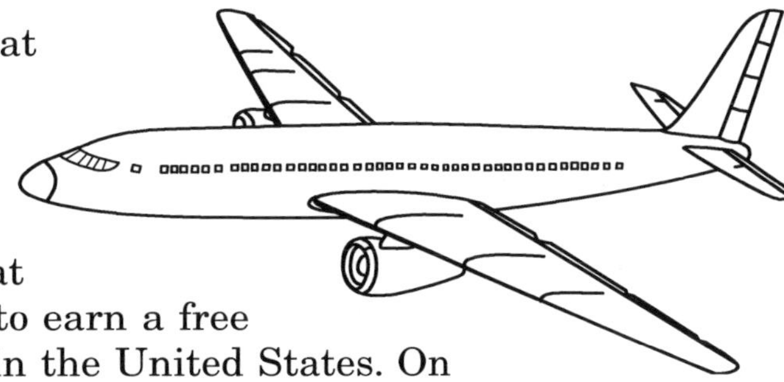

Solve

Step 1: Underline the sentence that tells you how many frequent-flyer miles Marius earns per dollar on his credit card.

He will earn one mile for every dollar he charges on his credit card.

Step 2: Write a proportion for dollars and miles. He must spend $25,000 to earn the trip.

$\frac{\$1}{1}$ frequent-flyer mile = $\frac{\$25{,}000}{25{,}000}$ frequent-flyer miles

Step 3: Now, find the number of years it will take to earn 25,000 frequent-flyer miles.

$\frac{\$25{,}000}{\$5{,}000} = 5$ years — Divide by the dollars he charges per year.

Step 4: Calculate the annual fees he will pay for this number of years.

$5 \times \$50 = \$250$

Answer the Question

Step 5: It will cost Marius $250 in annual fees to earn a free round-trip ticket, in addition to the $25,000 he spends over the 5 years.

Now try these problems.

1. Este gets a credit card that offers frequent-flyer miles. The annual fee is $80. She will earn two miles for every dollar she charges on her credit card. Her goal is to earn 16,000 miles for

a round trip she wants to take. On average, Este spends $4,000 in credit card charges per year. How much will it cost her in annual fees to earn this free round-trip ticket?

Answer: It will cost Este $________ in annual fees to earn a free roundtrip ticket.

2. Felicia gets a credit card that offers frequent-flyer miles. The annual fee is $40. She will earn one mile for every dollar she charges on her credit card. How much must she charge to earn 12,000 frequent flyer miles in one year? Besides annual fees, list factors that may make a free frequent-flyer ticket not quite "free."

Answer: ____

____

____

3. Some credit card companies offer protection plans. With a protection plan, your balance is paid off if you lose your job or have an accident and can't work. Corbin's credit card has a monthly fee for this service. The fee is 73 cents per $100 of unpaid balance. For the last 12 months, Corbin has maintained his unpaid balance at $1,000. How much has he paid for this service over the year?

A $7.30 **B** $87.60 **C** $120 **D** $876

4. Peter's credit card company sends him a check for $2.50 in the mail. When he cashes it, he will start earning 1 point for every credit card dollar he spends. One point is worth 1% of $1. He can start redeeming his points after 2,500 points. The maximum number of points he can accumulate is 60,000 points per year.

a. Circle the minimum dollar amount that he can redeem.

b. Draw a square around the maximum dollar amount that he can redeem in one year.

$2.50 $10 $25 $27.50 $250 $600 $2,500 $6,000

☆ *Challenge Problem*

You may want to talk this one over with a partner.

For every $2 you spend with your credit card, you get the same number of cents back. How much must you spend to get back $2?

Answer: $__________

Lesson 5 → Reading the Fine Print!

Example On March 21st, Savannah mails a $50 payment on her $1,205 balance to the credit card company. The closing date is the 20th of each month. Payments not received by that date will cause a late fee to be added to the account balance. The credit card company also charges 1.5% monthly interest on unpaid balances. What will Savannah owe after her $50 payment is received?

Unpaid Balance	Late Payment Fee
Up to $149.99	$20.00
$150 up to $1,199.99	$28.00
$1,200 and over	$40.00

Solve

Step 1: Determine whether Savannah will owe a late fee. Compare the date when Savannah pays her bill to the credit card closing date.

She mails $50 on March 21st. March 20th is the closing date. So she owes a late fee.

Step 2: Determine what late fee Savannah owes.

$1,200 and over	$40.00

Step 3: Calculate the interest on an unpaid balance of $1,205.

$1,205 × 1.5% = $18.08

Step 4: Calculate the new unpaid balance after the $50 payment is received.

$1,205 + $40 + $18.08 – $50 = $1,213.08

Answer the Question

Step 5: Savannah will owe $1,213.08 *after* her $50 payment is received.

Now try these problems.

1. On July 15th, Robb mails $10 of his $155 balance to the credit card company. The closing date is the 15th of each month. Payments not received by that date will cause a late fee to be added to the account balance. Late fees are $15 on balances up to $150; $25 on balances over $150. The credit card company also

charges 1.8% monthly interest on unpaid balances. Complete the table to show what Robb will owe after his $50 late payment is received.

Original Unpaid Balance	Late Payment Fee	Interest Owed	Payment	New Unpaid Balance

2. In the fine print, Ziggy reads that the over-the-credit-limit fee is $27.00 per month. He charges a weight-lifting machine that puts him over the top of his $1,500 credit limit. His next credit card payment is late, *but* it brings him below his limit. Circle the over-the-credit-limit fee that he owes.

   $0 $27 $54 $1,500 $1,527 $1,554

3. Jann is consolidating her debts. She makes balance transfers from three credit card accounts to her *OneStop* account. The fine print says that there is a transaction fee for each transfer. The fee is 3% of the amount of the transfer, and not less than $5 nor more than $50. Complete the table to show the transaction fee for each transfer.

Balance	$1,239.00	$1,770.00	$92.12
Transaction Fee			

4. Caris' credit card company sends him 12 "convenience checks." He doesn't see that the transaction fee for using a check is 5% of the amount of the check (and not less that $5). Of the three checks he writes, one is for $10.99. The other two are over $100 each. His transaction fee for all three is $20.45. What do the two other checks add to?

   A $200 **B** $210.99 **C** $309 **D** $409

☆ *Challenge Problem*

You may want to talk this one over with a partner.

You and a friend compare credit cards. One card has an annual fee of $80, a low interest rate, and several services (such as frequent-flyer airline mileage and discounts from numerous merchants). The other card has no annual fee, a higher interest rate, and only a few basic services (such as balance transfers). Which card suits you best, and why?

Unit 4 Review

Review What You Learned

In this unit you have used mathematics to solve many problems. You have used mental math and estimation, practiced basic operations, solved equations, and used statistics, ratios, and proportions.

These two pages give you a chance to review the mathematics you used and check your skills.

✔ Check Your Skills

1. Kathianne buys a used guitar for $245, and charges it on her credit card. At the end of the first month, she makes a payment of $25. The credit card company charges 1.9% monthly interest on unpaid balances. After Kathianne's $25 payment is deducted, interest is computed on the unpaid balance. Write an expression that shows how much Kathianne now owes.

Answer: ______________________________

If you need to review, return to lesson 1 (page 41).

2. Jamie charges her new surfboard on her credit card. The surfboard is on sale for $199. (The regular price is $245.) She makes no more purchases with her credit card. At the end of each month, she pays $25. The credit card company charges her 2.2% monthly interest on her unpaid balance. Complete the table to show what she owes at the end of each month.

End of month	1	2	3	4	5	6	7	8	9
Unpaid balance									

a. How long does it take her to pay?

b. Does she save money by buying the surfboard with her credit card? If so, how much?

Answer: a. It take Jamie __________ to pay.

b. She ______________ (saves/does not save) $________ by buying the surfboard with her credit card.

If you need to review, return to lesson 3 (page 45).

3. The closing date for Di's *Thistle* credit card is the 25th of each month. The closing date for her *FirstCharge* credit card is the 15th of each month. Each card offers a grace period of 30 days.

She makes a purchase on the 7th of the month. Which card allows her money to stay in the bank longer? Why?

A *Thistle;* payment due in 18 days

B *Thistle;* payment due in 48 days

C *FirstCharge;* payment due in 8 days

D *FirstCharge;* payment due in 38 days

If you need to review, return to lesson 3 (page 45).

4. Preston gets a credit card that offers frequent-flyer miles. The annual fee is $65. He will earn one mile for every dollar he charges on his credit card. He reads that it takes 20,000 miles to earn a free round-trip ticket anywhere in the United States. On average, Preston spends $2,500 in credit card charges per year. Write an expression to find the cost in annual fees alone to earn a round-trip ticket in the United States.

Answer: ______________________________ = $________

If you need to review, return to lesson 4 (page 47).

5. On November 15th, Tobias mails $100 of his $247.50 balance to the credit card company. The closing date is the 16th of each month. Payments not received by that date will cause a late fee to be added to the account balance. Late fees are $21 on balances up to $200 and $35 on balances over $200. The credit card company also charges 2.1% monthly interest on unpaid balances. Fill in the table to show how much Tobias owes in the two payment cases.

Payment	Original Unpaid Balance	Payment received by closing date	Late Payment Fee	Interest Owed	New Balance after payment
On-time	$247.50	Yes			
Late	$247.50	No			

If you need to review, return to lesson 5 (page 49).

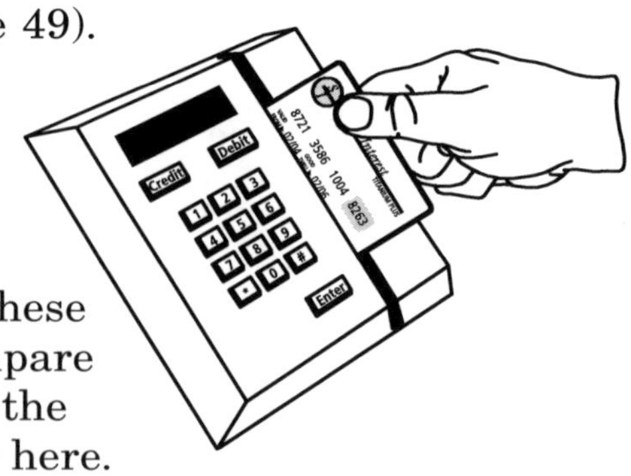

Write Your Own Problem

Choose a problem you liked from this unit. Write a similar problem using a situation and related facts from your own life. With a partner, share and solve these problems together. Discuss the mathematics and compare the steps you used. If you need to, rewrite or correct the problems. Write your edited problem and the answer here.

____

____

Unit 5 Rental Budgeting

Preview

How You Will Use This Unit

Rental budgeting involves many different things. You can rent almost anything: a house, a vehicle, tools or other equipment. You probably consider the length of time you rent, the cost of renting or buying, and the possibility of renting to own. As you compare options and make choices, you will often use math. The math skills you use include mental math and estimation, basic operations and equations, statistics, ratios, and proportions.

What You Will Do in This Unit

In this unit, math steps demonstrate how to solve problems. These steps can help you answer questions such as these:

You have a job in a new town. You rent an apartment for $525 per month. Your rent increases by 5% each year. What is the total amount of rent that you pay over five years?

You rent a car for one day while you are out of town. The company charges a daily rate plus a per-mile charge. The agent tells you the cost for 100 and 200 miles. Later, you wonder what the per-mile charge is, and what the daily rate is. What are these amounts?

You need a saw to cut down a tree in your yard. You find one that costs $189. You also find that you can rent one for $15 per day. How many uses make buying a better deal than renting?

The sticker price of a car is $17,499. You can lease it for $449 per month for two years. Then you have an option to buy it for $359 per month over three years. How much does this plan cost?

What You Can Learn from This Unit

When you complete this unit, you will have used mathematics to work problems related to rental budgeting. These problems are similar to those that may actually occur in your daily life.

Lesson 1

Houses & Apartments

Example Chuy gets a job as building inspector working for the city. He rents an apartment for \$754 per month. It is now five years later. His rent has increased by 5% each year since he moved in.

a. How much has Chuy paid in rent over these five years?

b. What is his average (mean) rent per month over this time?

Solve

Step 1: Calculate Chuy's rent for each year over the five-year period.

Year 1: \$754 per month
Year 2: \$754 × 105% = \$791.70 per month
Year 3: \$791.70 × 105% = \$831.29 per month
Year 4: \$831.29 × 105% = \$872.85 per month
Year 5: \$872.85 × 105% = \$916.49 per month

Step 2: Next, calculate the total amount he has paid over this period. (Remember to multiply by 12 for the number of monthly payments per year.)

(\$754 + \$791.70 + \$831.29 + \$872.85 + \$916.49) × 12 = \$49,995.96

Step 3: Now, find his average monthly rent over the five year period.

\$49,995.96 ÷ 60 = \$833.27 Divide by the number of months in 5 years.

Answer the Question

Step 4: **a.** Chuy has paid \$49,995.96 in rent over the five years.

b. His average rent per month over this time is \$833.27.

✏ Now try these problems.

1. Wendi gets a job as park ranger for the local state park. She rents an apartment for \$512 per month. Her rent increases by 7% each year over the four years that she lives there.

   a. How much does Wendi pay in rent over these four years?

   b. What is her average (mean) rent per month over this time?

   A \$2,273.25; \$454.65 **C** \$35,332.56; \$588.88

   B \$27,279.00; \$568.31 **D** \$73,740.84; \$1,536.27

2. Garth and Denise buy a house near the beach which they plan to rent to tenants. The realtor says their mortgage payments, plus utilities, will be about $1,145/month. She also says that they can rent the house for $1,200 per month during the six summer months, and $600 per month during the rest of the year. Suppose they rent the house out every month. Does the house pay for itself? If not, how much *more* money do Garth and Denise have to pay beyond the rent?

Answer: Rental house income = $__________ (summer) + $__________ (winter)

Total income from the house = $__________

Mortgage payments plus utilities for one year = $__________

The house ____________ (does/does not) pay for itself.

Garth and Denise have to pay $________ per year beyond the rent.

3. *YourMiniStore* quotes Glenn a fee of $75 per month for a 500 cubic-foot storage unit. The unit is 10 feet from floor to ceiling.

   a. Draw a diagram to show possible dimensions for the floor of the unit.

   b. How much will Glenn pay per square foot of floor?

Answer a.

b. Glenn will pay $______ per square foot.

☆ *Challenge Problem*

You may want to talk this one over with a partner.

You and your friend find an interesting problem in a book. Each letter stands for a digit. (Be careful, a letter may substitute for more than one digit!) What digit does each letter stand for? What code solves the problem? Explain your answer.

	3	F	F
	N	S	3
O	3	Z	E

____

____

____

Lesson 2

Renting Wheels

Example Ilene wants to rent a car for one day. The car rental agency charges a daily rate plus a per-mile charge. She asks how much it would cost to drive 100 miles, 160 miles, and 225 miles. The agent gives her the answers and she writes them in her notes. Later, her brother asks her what the per-mile charge is and what the daily rate is. What are these amounts?

Notes:
100 miles: \$41
160 miles: \$51.80
225 miles: \$63.50

Solve

Step 1: Underline the sentence that tells you how the rental company charges.

The car rental agency charges a daily rate plus a per-mile charge.

Step 2: Use x for the daily rate. Use y for the per-mile charge. Write an equation for the first line of data that Ilene wrote in her Notes.

$\$41 = x + 100y$ first equation

Step 3: Now, write an equation for the second line of data in Ilene's Notes. (You only need two equations to solve for two variables.)

$\$51.80 = x + 160y$ second equation

Step 4: Use these two equations to solve for x and y.

$x = 41 - 100y$ first equation (rewritten with x on the left)

$51.80 = (41 - 100y) + 160y$ Substitute this expression for x in second equation.

$51.80 = 41 + 60y$

$51.80 - 41 = 60y$

$10.80 = 60y$

$0.18 = y$

$x = 41 - 100(0.18)$ Substitute this value of y in first equation.

$x = 23$

Answer the Question

Step 5: The per-mile charge is \$0.18. The daily rate is \$23.

Now try these problems.

1. Mitchell rents an electric car for one day to tour the city. *Electri*City* charges a daily rate plus a per-mile charge. Mitchell asks how much

it would cost to drive 50 miles and 100 miles. The agent tells him $27.50 and $40. Later, he wonders what the per-mile charge is and what the daily rate is. Write two equations that you could solve together to find the answers.

Answer: ________________________ and ________________________

2. Jake's band calls ahead to rent a minibus. The company quotes $60 per 24-hours. The company says they won't charge for the next day if the bus is back less than an hour after it is due. The band has budgeted $240 for a minibus, and cannot afford any more. They pick the minibus up at 2 p.m. on Monday. What is the latest possible time by which they must deliver the minibus back? Mark the time on the calendar.

October 20	
20 Monday ***2 p.m. Pick up minibus***	**23** Thursday
21 Tuesday	**24** Friday
22 Wednesday	**25/26** Saturday/Sunday

3. Angelisa rents a car for two days. The company quotes a daily rate of $24.99 per day. When she gets the bill, the charge is $109.98. This is more than twice the daily rate. The company gives her three reasons. There is an $8 fee for being able to pick the car up on airport property. The rate goes up by $10 per day on weekends. They charged $32 to fill the gas tank. On which two days did Angelisa rent the car?
   - **A** Monday, Tuesday
   - **B** Wednesday, Thursday
   - **C** Friday, Saturday
   - **D** Saturday, Sunday

4. Brice and Ryan are going away to the same technical college. They each plan to rent a truck at $19.99 per day for two days to move their stuff. They find out that they will both move over the same weekend. So they rent the next largest truck for $29.99 per day and share the cost. Suppose each of them would have spent $39.98 moving independently. How much do they each save by sharing the larger truck for the same length of time?

Answer: $__________

☆ *Challenge Problem*

You may want to talk this one over with a partner. You and your family plus the family down the road are going on a trip. You can rent two cars or one minivan for yourselves and all your luggage. Which would you rent and why?

____

____

Lesson 3

Tools

Example Diego needs a power drill to help his mother do several things in her new home. He prices power drills and finds one that costs $79.99. He also finds a rental shop that rents power drills for $5 per day. How many times would Diego have to use the power drill to make buying a better deal than renting?

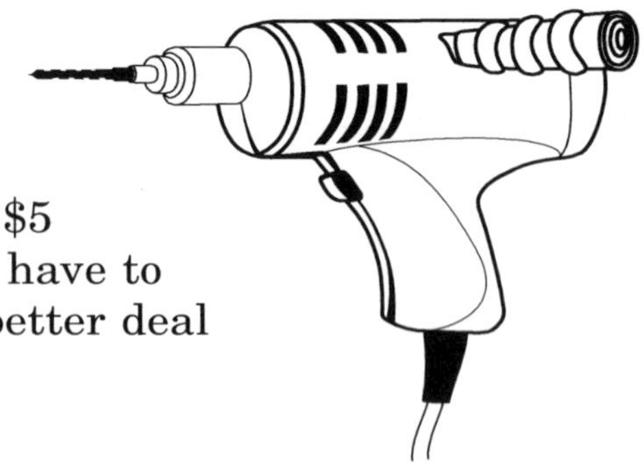

Solve

Step 1: Find how many rental days are contained in the purchase price.

$79.99 ÷ $5 = 15.998	Divide the cost of purchase by the rental fee.

Step 2: Now, write an inequality to show the number of days for which renting would be *more* than the cost of purchase.

16 × $5 > $79.99	Sixteen rentals at $5 each ($80) would cost more than buying.

Answer the Question

Step 3: Diego would have to use the power drill for 16 days to make buying a better deal than renting.

Now try these problems.

1. Marci and Luis, with their six-month old daughter, visit Marci's grandparents. At *Second*Time*Around,* Marci's grandmother finds a crib and a high chair for $178. Marci says it would be cheaper to rent these items at $15 a weekend. How many weekend visits would make buying a better deal than renting?

   Answer: ________ weekend visits would make buying a better deal than renting.

2. Dane finds a rental shop that rents brush chippers at $150 per day. He asks his neighbors if they would be interested in sharing the rent for a day. Out of twelve neighbors, five say *Yes,* two say *Maybe,* and the others say *No*. What is the per-person rental cost if Dane can convince the neighbors who said *Maybe* to say *Yes*?

   A $12.50 **B** $15 **C** $18.75 **D** $25

3. Hilde buys a party tent for $650. She rents it at $20 a time to neighbors who want to hold outdoor parties. She wants to know how quickly she can start to make money by renting out the tent.

   a. Fill in the table to show how quickly Hilde can pay off the tent.
   b. Circle the number of times she must rent it per month to start making money within the year.

Number of times per month that Hilde rents out the tent	Number of months before she starts making money
1	
2	
3	
4	

4. Mihaly rents equipment to clean the carpet and repaint his apartment before moving. He wants to get his deposit of $285 back. The rental cost of the equipment adds to $9 per hour. He picks it up at 8 a.m. and delivers it back at 4 p.m. The materials cost him $45. His time is worth $11 per hour. In your opinion, was it worth the effort? Explain.

   Answer: ____

   ____

☆ *Challenge Problem*

You may want to talk this one over with a partner.

You paint all the outside surfaces of this 3 × 3 × 3 cube made of small cubes. Do any of the small cubes have *no* painted surfaces? If so, how many? __________ Fill in the table to show the number of cubes for each number of painted surfaces.

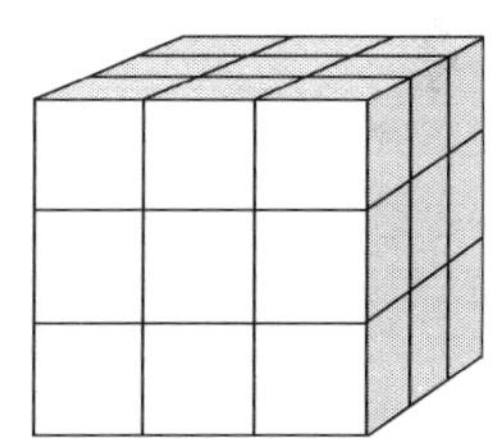

# of painted surfaces on a small cube	0	1	2	3	4	5	6
# of cubes							

What is the total number of small cubes in the large cube? ____________
Check that the total of your second row in the table equals this number.

Lesson 4 Chain of Rentals

♠ A Card Game (for Two or More Players)

The goal of this game is to make a chain of rental domino cards. First, a player puts down a domino card. Then players add other domino cards to continue the chain.

Materials

Rental domino cards (on the next page), two number cubes

Directions

1. Place the domino cards face down on the playing area. Each player picks up seven cards. Players toss the number cubes until one player tosses a double. This player becomes player 1 and starts the game. Player 1 places a domino card, face up, on the table. Player 1 then picks up a domino card from the remaining face down cards to replace the one played.

2. Player 2 examines cards for a match in the chain. A match occurs when one of these four things happens:
   - A card matches the dollars on an open end of the rental chain. (For example, 5 days at $20 results in, or matches, $100.)
   - You can make the dollar value on a card match the result of a calculation with the time and dollars on an open end of the rental chain. (For example, $4 can be the daily rate if you pay $20 over 5 days. Or 5 days at $20 per day could be a total of $100.)
   - The time period on a card matches the time period in the rental chain. (For example, one month matches four weeks.)
   - A wild card (*) can be used anywhere because it matches anything.

3. Player 2 puts down a card, at right angles to, or in line with, one of the open ends. Player 2 then picks up a card from the remaining cards to replace the one used. If player 2 cannot make a play, then that player misses a turn.

4. Players take turns placing cards. When all the cards are used, players continue with the cards in their hands until one player has no more cards. The winner is the first player without any cards.

Before you play the game, try this warm-up problem.

1. You have these cards. 2 hours — $5; 5 hours — $4; 10 hours — $1; 2 weeks — $70; 4 weeks — $140; 4 weeks — $2,048; 3 months — $210. Which card can you use on the sample *Chain of Rentals*?

   Answer: ______________________________

Sample *Chain of Rentals*

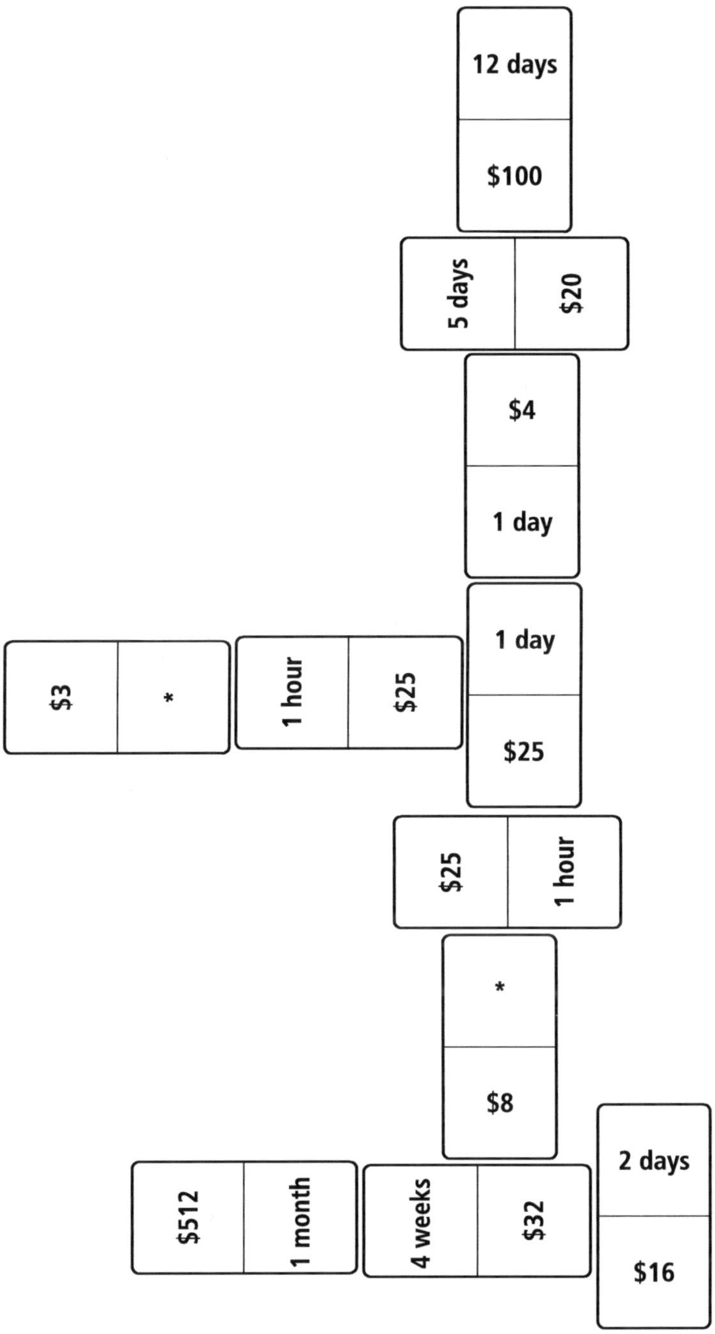

Rental domino cards:

Make cards with the following time / dollar combinations:

1 hour — $3	1 day — $5
1 hour — $5	1 day — $4
1 hour — $25	1 day — $15
2 hours — $5	1 day — $25
5 hours — $4	1 day — $35
5 hours — $25	2 days — $16
10 hours — $1	3 days — $48
24 hours — $15	5 days — $20
36 hours — $40	12 days — $100
48 hours — $30	15 days — $360
72 hours — $20	32 days — $16
1 week — $105	2 weeks — $70
1 week — $128	3 weeks — $75
1 week — $245	4 weeks — $32
1 week — $360	4 weeks — $2,048
2 weeks — $32	4 weeks — $140
1 month — $512	2 months — $720
1 month — $1,220	2 months — $50
1 month — $1,440	3 months — $210

Make these wild cards:

* — $3	* — 1 hour
* — $8	* — 1 day
* — $100	* — 1 week
* — $360	* — 1 month
* — $1,200	* — *

You can make more rental domino cards to extend the game.

Lesson 5

Renting to Buy

Example Aria wonders whether to buy or lease a new car. The sticker price of the car is $16,893. She can lease the car for $369 per month for two years. At the end of the lease, she can then buy the car, and pay $435.79 per month for three years.

a. How much does this plan cost?

b. How much *more* does the plan cost than the sticker price?

Solve

Step 1: Find the cost of leasing the car for two years.

$369 × 24 = $8,856 — Multiply cost per month by 24 months.

Step 2: Next, find the cost of buying the car at the end of the lease.

$435.79 × 36 = $15,688.44 — Multiply cost per month by 36 months.

Step 3: Now, add to find the total cost of the plan. Then, calculate the difference between the cost of the plan and the sticker price.

$8,856 + $15,688.44 = $24,544.44

$24,544.44 – $16,893 = $7,651.44

Answer the Question

Step 4: **a.** The plan costs $24,544.44.
b. This is $7,651.44 more than the sticker price.

Now try these problems.

1. Zak wonders whether to buy or lease a new car. The sticker price of the car is $20,499. He can lease the car for $405 per month for two years. At the end of the lease, he can buy the car, and pay $452.59 per month for three years. Write an expression and find the difference between the plan cost and the sticker price.

   Answer: ______________________________ = __________

2. Cason can lease a car or he can buy it outright. On the lease plan, he will pay $295 per month for two years, then buy the car for $12,540. On the purchase plan, he will pay $464.56 per month

over four years. How much does each plan cost? Which plan do you prefer and why?

Answer: Lease plan costs $__________.

Purchase plan costs $__________.

____

____

3. Kota and Jeff put down $7,500 on a $75,000, old Mediterranean-style home. Their mortgage payment is $1,075 per month. They live in one part of the house. They rent out three other sections of the house for $475 per month each. They are saving for a $15,000 down payment on their dream home. Suppose nothing changes. How long will it take them to save the down payment from this rental income (after they use the rental income to pay the mortgage)?

A About 1 year

B About $2\frac{1}{2}$ years

C About $3\frac{1}{2}$ years

D Exactly 5 years

4. The rent-to-own business grew from 2,000 to 7,000 stores in one decade.

   a. What was the mean rate of growth per year?

   b. Suppose this rate of growth continued. How many stores would there be in five more years? Draw a diagram to show this data.

Answer: **a.** The mean rate of growth over the decade was ________ stores per year.

b.

☆ *Challenge Problem*

You may want to talk this one over with a partner.

In a certain arithmetic sequence, the value of (3 × third term) is equal to the value of (6 × sixth term). What is a simple arithmetic sequence that satisfies this condition? In your sequence, what is the value of (9 × ninth term)?

____

____

Unit 5 Review

Review What You Learned

In this unit you have used mathematics to solve many problems. You have used mental math and estimation, practiced basic operations, solved equations, and used statistics, ratios, and proportions.

These two pages give you a chance to review the mathematics you used and check your skills.

✔ Check Your Skills

1. Sasha gets a job as a teacher's aide at the local high school. He rents an apartment for $625 per month. It is now three years later. His rent has increased by 10% each year since he moved in.

a. How much is Sasha's rent per month in each of these three years?

Answer: Year 1 – Sasha's rent is $_________ per month.

Year 2 – Sasha's rent is $_________ per month.

Year 3 – Sasha's rent is $_________ per month.

b. What is the total amount that Sasha pays in rent over this three year period?

Answer: $_________

c. What is his average (mean) rent per month over this time?

Answer: $_________

If you need to review, return to lesson 1 (page 54).

2. Dessa rents a car for one day while she is out of town at a conference. *TodaysWheels* charges a daily rate plus a per-mile charge. Dessa asks how much it would cost to drive 150 miles and 250 miles. The agent tells her $50 and $60. Later, she wonders what the per-mile charge is and what the daily rate is. What are these figures?

Answer: The per-mile charge is $_________.

The daily rate is $_________.

If you need to review, return to lesson 2 (page 56).

3. Aspen needs a sewing machine to sew curtains for her new apartment. She prices sewing machines and finds one that costs $99 on sale. She also finds a rental shop that rents sewing machines for $15 per day. How many days would Aspen have to use the sewing machine to make buying a better deal than renting? Explain how to find this number.

Answer: ____

____

If you need to review, return to lesson 3 (page 58).

4. You wonder whether to buy or lease a new car. The sticker price of the car is $17,499. You can lease the car for $449 per month for two years. At the end of the lease, you can buy the car, and pay $359 per month for three years. How much *more* does the plan cost than the sticker price?

   A $6,201 **C** $7,401

   B $7,281 **D** $23,700

If you need to review, return to lesson 5 (page 62).

Write Your Own Problem ✍

Choose a problem you liked from this unit. Write a similar problem using a situation and related facts from your own life. With a partner, share and solve these problems together. Discuss the mathematics and compare the steps you used. If you need to, rewrite or correct the problems. Write your edited problem and the answer here.

____

____

____

____

____

____

____

____

Unit 6 Self-Employment

Preview

How You Will Use This Unit

Self-employment involves many different things. Starting up and managing your income and expenses are two examples. You probably also consider insurance costs and taxes. You may also think about getting a loan. As you compare options and make choices, you will often use math. The math skills you use include mental math and estimation, basic operations and equations, statistics, and ratios and proportions.

What You Will Do in This Unit

In this unit, math steps demonstrate how to solve problems. These steps can help you answer questions such as these:

You budget $100 to have business cards printed for your catering business. A local print shop gives you a quote of $75 for a box of 500. There is an additional charge of $10 for every 100 cards over 500. How many cards can you have printed?

You make and sell a mix to make a dip for chips. For every dollar you spend to make your chip mix, only 20 cents goes toward the ingredients. You charge $4 for a packet of mix that costs you $2 to make. You charge $6 for a packet that is 50% larger. How much profit do you make?

You join an association for self-employed people. You used to pay $195 per month for health insurance. Now you pay $167 per month. How much do you save per year?

You have a line of credit at your bank. During the low season, you borrow $4,722 against this line of credit. During the high season you pay the loan back. What is the mean extra income that you must make per month during the high season to pay back this loan?

What You Can Learn from This Unit

When you complete this unit, you will have used mathematics to work problems related to self-employment. These problems are similar to those that may actually occur in your daily life.

Example Franco budgets $100 to have business cards printed for his delivery business. A local print shop gives Franco a quote of $59 for a box of 500 cards in two colors. This charge includes a $9 setup charge. There is an additional charge of $12 for each set of 100 cards over 500. (The shop only prints cards in blocks of 100.) How many cards can Franco have printed?

Solve

Step 1: First, subtract the quote for 500 cards from Franco's budgeted money.

$100 – $59 = $41

Step 2: Next, calculate how many additional sets of 100 cards Franco can afford.

$41 ÷ $12 = 3.42

3 sets — The print shop only prints cards in blocks of 100.

Step 3: Now, calculate the total number of cards that Franco can have printed.

500 + 300 = 800 — 500 for $59; 300 at an additional $12 per 100 over 500

Answer the Question

Step 4: Franco can have 800 cards printed.

Now try these problems.

1. Pia budgets $500 to have brochures printed. A local print shop gives her a quote of $350 for 1,000 brochures. This charge includes a $10 setup charge. There is an additional charge of $20 for each set of 50 brochures over 1,000. How many brochures can Pia have printed?

   Answer: Pia can have ____________ brochures printed.

2. Teryn is trying to keep the start-up costs down for her landscaping business. She wants to print 200 color flyers to pass around the neighborhood. A local printer gives her a quote of $45

to print 50 color flyers. The cartridges for her color printer she would need cost $79.50. A ream of 500 sheets of paper costs $12.99. What is the amount of money she might save by printing the flyers herself? What are your assumptions?

Answer: Cost of 200 flyers printed by local printer =

____

Cost of supplies for printing the flyers herself =

____

Amount of money she saves = $__________

Assumptions: ______________________________

3. Mari's start-up costs include setting up an office. Her uncle loans her $230 to buy an old desk and chair, $1,205 for a computer, $565 for office equipment, and $250 for supplies. Mari agrees to pay back the loan in equal monthly installments over one year. How much should she budget per month to cover these expenses?

   A $46.88 **C** $562.50

   B $187.50 **D** $2,250

4. Lark plans to take a one-week vacation. She wants to get ahead on her work hours before she goes. On average, over four weeks, she works 150 hours. Her goal is to work the same number of hours during the four-week period that includes her vacation. How many *more* hours must she work per week for each of the three weeks to meet her goal?

   Answer: ________ hours

☆ *Challenge Problem*

You may want to talk this one over with a partner.

You decide that a color printer is the most important part of your graphic arts system. You find exactly the right printer for $350. But you only have $300. You put the money into a savings account with a 6% annual interest rate, compounded monthly. About how much money will be in the account at the end of a year? Explain your process.

____

____

____

Lesson 2 Income versus Expenses

Example Waverly makes and sells bread pudding. Out of every dollar she spends to make the pudding, 20 cents pays for the ingredients. The rest covers the electricity, insurance, purchasing, and advertising. Waverly charges \$7 for a slice of bread pudding that costs her \$5 to make. She charges \$10.50 for a portion that is 50% larger. The non-food expenses for the larger portion are the same as for the smaller slice.

a. How much do the *extra* ingredients cost?

b. How much profit does Waverly make when she sells the larger portion of bread pudding?

Solve

Step 1: Underline the sentence that tells how much money goes toward ingredients.

Out of every dollar she spends ..., 20 cents pays for the ingredients.

Step 2: Underline the sentence that tells the cost to make a \$7 slice of bread pudding.

Waverly charges \$7 for a slice of bread pudding that costs her \$5 to make.

Step 3: Use a proportion to find the cost of the ingredients for a \$7 slice of pudding. Use x to represent that cost.

$\frac{\$0.20}{\$1.00} = \frac{x}{\$5}$

So $x = \$1$ The ingredients for a slice cost \$1.

Step 4: Next, calculate the cost of the *extra* ingredients for a \$10.50 portion of pudding.

$\$1 \times \frac{1}{2} = \$0.50$ A portion that is 50% larger uses 50% more ingredients.

Step 5: Now, calculate the profit that Waverly makes on the \$10.50 portion of pudding.

$\$10.50 - (\$5 + \$0.50) = \$5$ Remember that the non-food expenses stay constant.

Answer the Question

Step 6: **a.** The extra ingredients cost \$0.50.
b. Waverly makes \$5 profit when she sells the larger portion.

Now try these problems.

1. Out of every dollar Cameron spends to make a special mix for a chip dip, he spends 20 cents on ingredients. The rest pays for electricity, purchasing, packaging, and advertising. Cameron charges $3 for an 8-ounce packet of mix, which costs him $2 to make. He charges $4.50 for a packet that is 50% larger. The non-food expenses stay constant.
   a. How much do the *extra* ingredients cost?
   b. Make a table to show his profit for larger and larger packets of his special mix.
   c. What assumption(s) do you make?

   Answers: a. The extra ingredients cost $__________.

   b.

Packet size	Sale price	Cost to make mix	Profit
8 ounces	$3	$2	
12 ounces	$4.50		
16 ounces	$6		
20 ounces	$7.50		

   c. Assumption(s): ______________________________

2. Saharah sets up a fruit stand on two opposite corners of an intersection. Two other fruit stand operators set up one fruit stand each on the other two corners. Which math does Saharah use to explain that she should get more business than the other two operators? Explain.

   A geometry **C** probability
   B algebra **D** measurement

   Explanation: ______________________________

3. Trista's band makes a CD. They earn 1.5% royalties on the sale of the CD. The band earns $4,829.10 in royalties. Suppose the CDs sell for $12.50 each. About how many CDs have been sold?

   Answer: ______________________________

☆ *Challenge Problem*

You may want to talk this one over with a partner.

$

The vertical line in a dollar sign separates the **S** into four segments. You draw 100 vertical lines close to each other through the **S**. How many segments have you separated the **S** into? How do you figure this out? Write an expression for the number of segments.

____

____

Lesson 3 — Insurance & Taxes

Example Serena joins an association for people who are self-employed. Annual membership fees for the association are $90. Before joining, she paid $270 per month for health insurance. Now she pays $195 per month for the same coverage through the association.

a. How much money does she save per year, after paying her membership fee?

b. She has a deductible of $2,000. What is the most that she will pay in health costs in one year? Include the fee and monthly insurance premiums in her total costs.

Solve

Step 1: Underline the sentences that tell you how much Serena paid per month before, and then through the association.

Before joining the association, she paid $270 per month for health insurance. Now she pays $195 per month … through the association.

Step 2: Find the difference in payments per month.

$270 – $195 = $75

Step 3: Now, calculate how much she saves per year, after she pays her fee.

($75 × 12) – $90 = $810 She saved $75 per month. She pays a $90 fee.

Step 4: Suppose Serena has over $2,000 in medical bills. Write the maximum amount of these bills that she will pay. Add insurance premiums and member's fee.

$2,000 + ($195 × 12) + $90 = $4,430

Deductible + premiums + member's fee

Answer the Question

Step 5: **a.** Serena saves $810 per year, after her member's fee.
b. The maximum amount that she will pay in health costs in one year is $4,430.

Now try these problems.

1. Jon joins an association for self-employed people. The cost of membership is $120 per year. He used to pay $275 per month for health insurance with another organization. Now he pays $220 per

month for the same coverage, through the association. How much money does he save per year, after paying his membership fee?

A \$540 **B** \$660 **C** \$780 **D** \$2,100

2. Chad owns a fitness consulting business. He pays \$215 per month for health insurance premiums. Since he is self-employed, he cannot deduct these premiums as a business expense. Instead, he pays an additional 15.3% tax on the premiums. Write an expression for the amount of tax that Chad pays per year because of this. Find the amount.

Answer: ______________________________ = __________

3. Dustin is a carpenter. Vendors and customers visit him in his workshop. He pays \$50 per year for \$2,000,000 liability insurance. This protects him in case visitors slip or hurt themselves in his shop. His deductible is \$3,000.
   a. How much does he pay per year in insurance premiums?
   b. What is the maximum amount of money that he might pay in one year?

Answer: a. \$________ **b.** \$____________

4. Kenetti runs a babysitting service. In her first year of operation, her gross income is about \$13,500. Of this, she pays 15% federal income tax, 5% state income tax, and 6.1% social security tax. Circle the column that shows what she pays for each of these taxes.

Federal Income Tax	\$202.50	\$2,025	\$20,250
State Income Tax	\$675	\$675	\$1,012.50
Social Security	\$1,006.50	\$823.50	\$1,006.50

☆ *Challenge Problem*

You may want to talk this one over with a partner.

Look at this sequence of figures. How many dots will be in the *fifth* figure of the sequence? What do you have to do to create each figure from the previous figure?

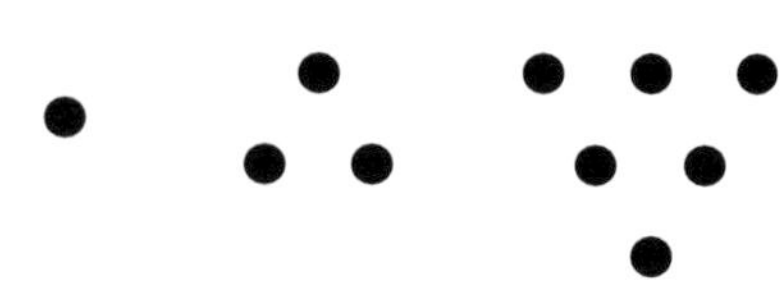

____

____

____

____

Lesson 4 — Getting Loans

Example Miguel has a line of credit at his bank. His credit limit is $5,000. During the low season, he writes three checks against his line of credit to cover operating expenses. During the four months of the high season, he pays the loan back in equal installments. What is the mean (average) *extra* income that he must make per month during the high season to pay back the loan?

Checks written against bank loan:

Rent:	$1,050
Utilities:	$875
Supplies:	$521

Solve

Step 1: First, add the checks written against the bank loan.
$1,050 + $875 + $521 = $2,446 This is the amount Miguel must pay back.

Step 2: Underline the sentence that tells the number of months to pay back the loan.
<u>During the four months of the high season, he pays the loan back in equal installments.</u>

Step 3: Now, calculate the amount of each monthly installment.
$2,446 ÷ 4 = $611.50

Answer the Question

Step 4: The mean (average) *extra* income that Miguel must make per month during the high season to pay back the loan is $611.50.

Now try these problems.

1. Cammie has a line of credit at her bank. Her credit limit is $6,000. During the low season, she writes four checks against her loan, to cover operating expenses. The checks are for $435, $291, $719, and $103. During the six-months of the high season, she pays the loan back in equal installments. What is the mean *extra* income that she must make per month to pay back the loan?

   Answer: Cammie must make an extra $__________ per month during the high season.

2. Jesicanne is a graphic artist. Before she starts a project, she asks for payment of half the fee. When a customer agrees to her

design, she is paid another quarter of the project fee. Today a customer agreed to the design she made, and paid her $525. What percent of the total project fee is not yet paid?

A 0% **B** 25% **C** 50% **D** 75%

3. Luis gets a long-term loan of $12,000 to finance new sound-recording equipment. The fixed annual interest rate is 2%, and the repayment term is 5 years. Each year he pays back one fifth of the original loan plus interest on the unpaid balance. How much does Luis pay each year over the period of this loan? Complete the table to show his yearly installment and the interest paid at the end of each year.

End of year	Installment	Interest
1		
2		
3		
4		
5		

4. Through the Small Business Administration, Gaea gets a loan of $9,450 for her new gardening business. She agrees to pay the principal back in equal installments. The monthly interest rate (simple interest) is 1.5%. The term of the loan is one year.

a. How much principal does she pay each month? How much interest does she pay each month?

b. How much does Gaea pay in total for her loan, including principal and interest?

Answer: a. ____________________

b. __________

☆ *Challenge Problem*

You may want to talk this one over with a partner.

You have a small business. At times, your income does not cover your expenses. Before going to your bank for a loan, you look for ways to improve your cash flow. What are some ways in which you might do this?

____

____

Lesson 5 — Making the Numbers Work

♠ A Card Game (for Two or More Players)

The goal of this game is to combine five numbers to get a specified number. You look at the numbers on the cards in your hand. Then you insert operations and grouping symbols to get the specified number. Your expression must use all of the cards in your hand.

Materials

Card deck for *Making the Numbers Work* (on the next page), paper and pencil for each player

Directions

1. Shuffle the cards and deal five cards to each player. Place the next card face up on the the table. Place the rest of the card deck face down, in the center of the table. Players sit around the table.
2. Players look at the cards in their hands. You can use any operation, including +, –, ×, ÷ to combine your five cards. You can also use grouping symbols such as ( ), [ ], and { }, to create expressions. The goal is to use all five of your cards in one expression to equal the number in the center. For example, the five numbers, 3, 5, 6, 7, and 10, can be combined to equal 2 by writing them like this: $(7 - 6) + [3 - (10 \div 5)]$.
3. The first player to write an expression that equates to the number on the card in the center of the table wins 3 points. The second and third players who create *different* expressions that also equal that number win 2 points and 1 point. Then put all the players' cards and the card in the center of the table in a discard stack.
4. Deal five more cards to each player, and place the next card face up, in the center of the table. When the deck is all dealt, replace it with the shuffled discard stack. The game goes on until one player has 15 points, or as long as the players choose to continue. The winner is the player with the most points.

✏ Before you play the game, try these warm-up problems.

1. Tracey has 3, 10, 19, 21, and 24 in her hand. What expression can she make that equals 20?

   Answer: ______________________________

2. Jerome has 2, 5, 7, 10, and 18. What expression can he make that equals 20?

   Answer: ______________________________

Card Deck for *Making the Numbers Work*

Make 3 cards for each of the numbers 1 through 10.
Make 2 cards for each of the numbers 11 through 17.
Make one card for each of the numbers 18 through 25.

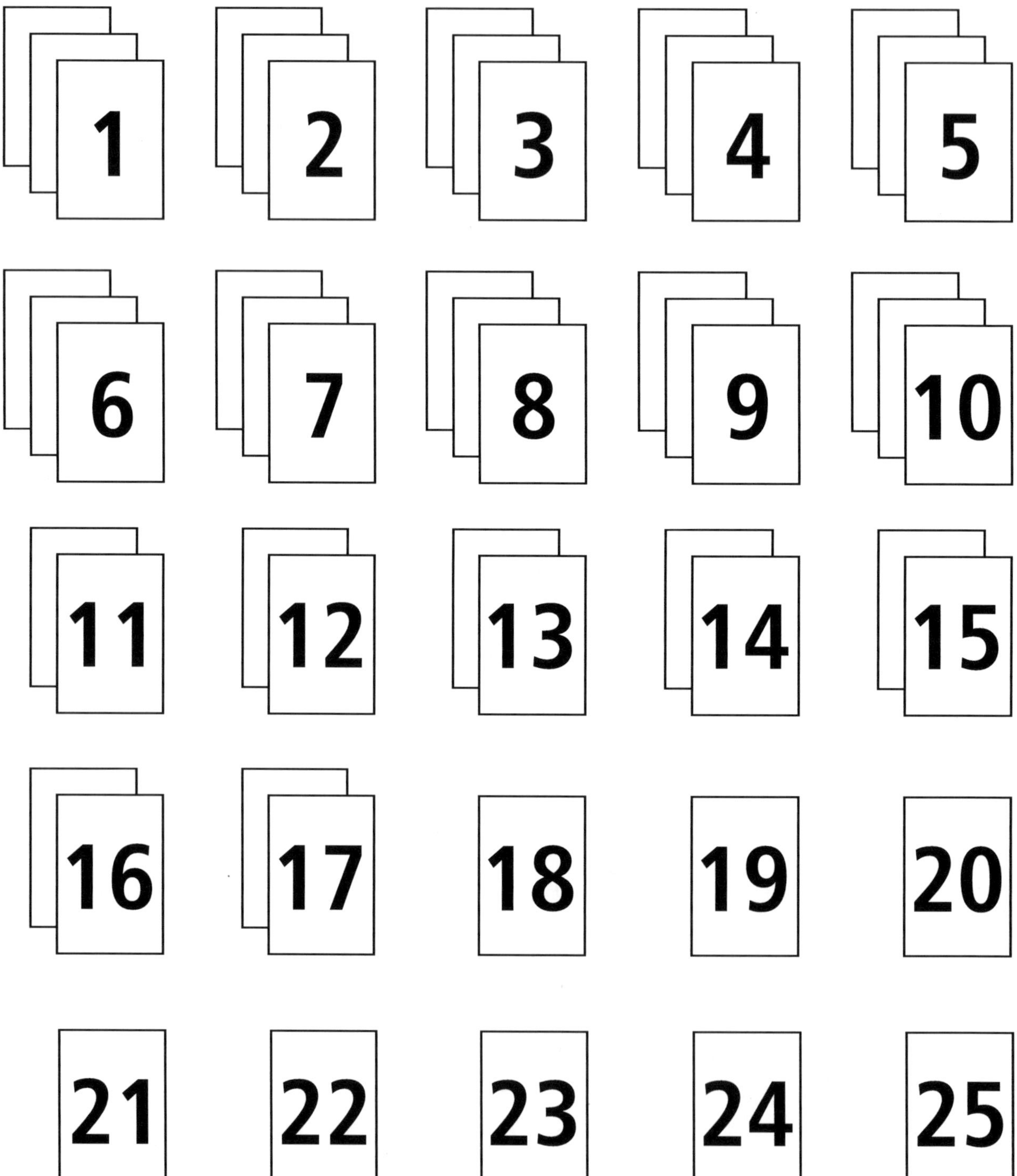

Unit 6 Review

Review What You Learned

In this unit you have used mathematics to solve many problems. You have used mental math and estimation, practiced basic operations, solved equations, and used statistics, ratios, and proportions.

These two pages give you a chance to review the mathematics you used and check your skills.

✔ Check Your Skills

1. Darryl budgets $100 to have flyers printed for his fishing guide. A local print shop charges him $85 for 100 flyers. This charge includes a $7 setup charge. There is an additional charge of $8 for every 10 flyers over 100.
   - **a.** How many flyers can Darryl have printed?
   - **b.** How much *more* money would Darryl need to get one more set of 10 flyers?

   Answer: a. ________ **b.** ________

   If you need to review, return to lesson 1 (page 67).

2. Craig mixes and sells packets of special healthy trail mix. For every dollar he spends to make his trail mix, only 20 cents goes toward the ingredients. The rest pays for electricity, purchasing, packaging, and advertising. Craig charges $4 for an average packet of trail mix, which costs him $2 to produce. He charges $6 for a packet that is 50% larger. The non-food expenses stay constant.
   - **a.** How much do the extra ingredients cost?

     A $0.20 **B** $0.40 **C** $0.50 **D** $1.00
   - **b.** Write an equation to give his profit on the $6 packet of trail mix. Find the amount.

   Answer: b. An equation that gives his profit is

   ________________________________ = __________

   If you need to review, return to lesson 2 (page 69).

3. Loren joins an association for self-employed people. The cost of membership is $95 per year. She used to pay $180 per month for

health insurance with another organization. Now she pays $155 per month for the same coverage, through the association.

a. How much money does she save per year, after paying her membership fee?

b. Her deductible is $2,000. What is the maximum amount that she might pay in one year? Circle the entry in the table that shows the correct amounts.

Money saved per year	Maximum amount that she will pay in one year for health costs
$205	$3,955
$300	$2,300
$395	$1,605

If you need to review, return to lesson 3 (page 71).

4. Cason has a line of credit at his bank. His credit limit is $4,000. During the low season, he borrows $3,840 against this line of credit to cover operating expenses. During the four months of the high season, he pays the loan back in equal installments. What is the mean (average) *extra* income that he has to make per month during the high season to pay back this loan?

A $320 **B** $640 **C** $960 **D** $1,920

If you need to review, return to lesson 4 (page 73).

Write Your Own Problem

Choose a problem you liked from this unit. Write a similar problem using a situation and related facts from your own life. With a partner, share and solve these problems together. Discuss the mathematics and compare the steps you used. If you need to, rewrite or correct the problems. Write your edited problem and the answer here.

Unit 7 — Going on a Trip

Preview

How You Will Use This Unit

Going on a trip involves many different things. Making reservations and renting special equipment are two examples. You probably also consider sources for cash, such as ATMs. You may also think about the impact on your budget. As you compare options and make choices, you will often use math. The math skills you use include mental math and estimation, basic operations and equations, statistics, and ratios and proportions.

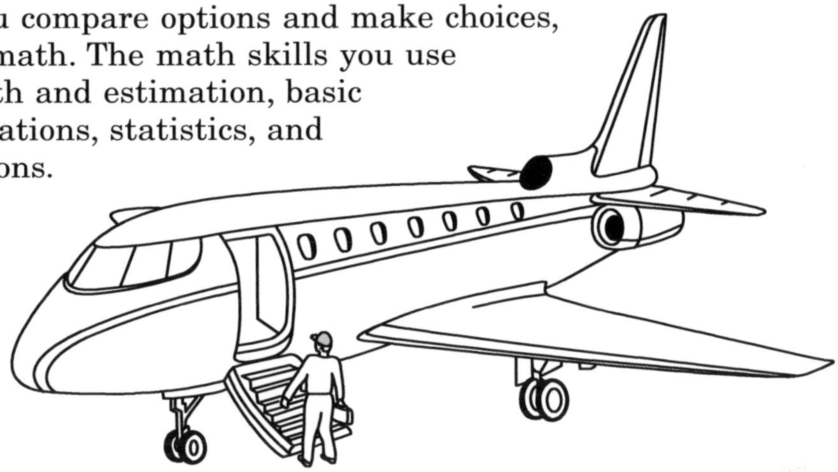

What You Will Do in This Unit

In this unit, math steps demonstrate how to solve problems. These steps can help you answer questions such as these:

You plan a 10-day cruise. The cruise costs $635. The round-trip plane ticket to get to the cruise ship costs $152. You spread the cost over 12 months. What is the monthly cost?

You have a ticket for flight 91. The ticket costs $250. Usually 4% of the ticketed passengers don't show up. The airline offers you a $200 voucher to give up your seat because the airline overbooked its seats. What percent of a future trip, at the same price, do you save?

You rent ski equipment for your week of skiing. The rent is $305. If you take your own equipment, the airline tickets will only cost $40 more each way. What is the difference in cost?

Today, the exchange rate for dollars to English pounds sterling is $1.6272 to £1. *How many pounds sterling can you get for $250?*

What You Can Learn from This Unit

When you complete this unit, you will have used mathematics to work problems related to going on a trip. These problems are similar to those that may actually occur in your daily life.

Lesson 1

Budget Impacts

Example Tasha is planning a 7-day, 6-night vacation. Her round-trip plane ticket costs $756. Her hotel room costs $97 per night. She allows $150 for items that she would not buy at home. Suppose she spreads the cost of her vacation equally over 12 months. What is the effect on her budget each month?

Solve

Step 1: Find the cost of Tasha's hotel room.

$97 × 6 = $582 — She plans to stay for 6 nights.

Step 2: Next, find the total cost of the vacation.

$582 + $756 + $150 = $1,488 — Add the room, ticket, and other items.

Step 3: Now, find the cost per month over 12 months.

$1,495 ÷ 12 = $124

Answer the Question

Step 4: The effect on Tasha's budget is to increase expenses by $124 per month.

Now try these problems.

1. Aaron plans a 14-day, 13-night vacation. The hotel costs $85 per night. The round-trip plane ticket costs $234. He allows $500 for other items. Suppose he spreads the cost of the vacation equally over 6 months. What is the effect on his budget each month?

Answer: The effect on Aaron's budget is ________________________.

2. Canda puts the cost of the hotel on her credit card so she can delay paying the cost. The hotel costs $399.54. The credit card company charges 1.8% monthly interest on unpaid balances. How much does she owe at the end of the month?

A $399.54 **C** $471.46
B $406.73 **D** $1,016.46

3. Ankit plans to go away for a month's vacation. He looks at his average expenses over the past six months. He will leave his car behind, but he will maintain his apartment. About how much (to the nearest $50) of his typical expenses can he safely apply to his month's vacation? Explain.

Average Expenses over last 6 months

Rent	$465
Utilities	$125
Telephone	$55
Cable TV	$25
Car payment	$205
Gas for car	$110
Food	$350
Clothes	$75
Entertainment	$60

Answer: ____

____

4. Shonda looks for ways to save money for her trip. She does not take her car. Her auto insurance company allows her to suspend the collision and liability parts of her insurance. They tell her she will save 3% on her $1,200 premium. Suppose each square represents $\frac{1}{100}$ of her auto insurance premium.

a. Shade the squares that represent 3%.

b. Shonda has $____________ extra for her trip.

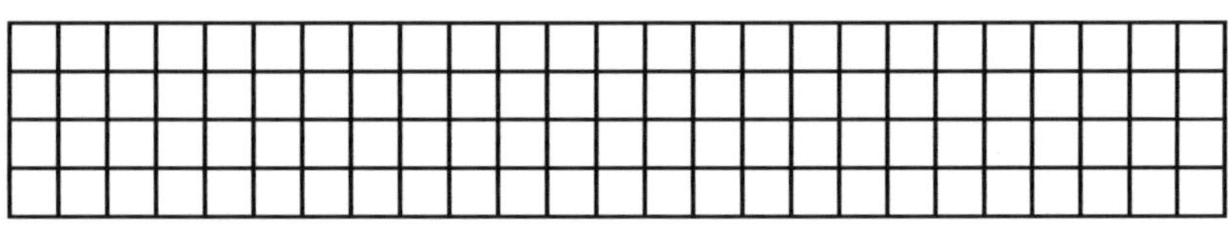

☆ *Challenge Problem*

You may want to talk this one over with a partner.

You and a friend need to save $200 each for a ski trip over the holidays. You start with $46 and get a job that pays $7 per hour. Your friend starts with $21.50, and gets a job that pays $8.50 per hour. Write an expression for the number of hours that each of you must work to save $200. Who has to work more hours, and by how much?

Answer: You have to work ____________________ hours.

Your friend has to work ____________________ hours.

____

____

____

Lesson 2 — Currency Translations

♠ A Card Game (for Two or More Players)

The goal of this game is to translate numbers between one system and another. First, you pick a card. Then you translate and check.

Materials

Currency Translation cards (starter deck on the next page), paper and pencils

Directions

1. Shuffle cards, and put the deck face down in the center of the table. Give each player paper and pencil to write their answers. Players sit around the table.
2. Player 1 takes the top card of the deck. Place it face up on the table. Player 1 then names a base between 2 and 10, other than the base on the card. For example, 12_5 names a number in base 5. The number 12_5 means $(1 \times 5^1) + (2 \times 5^0)$. This number is equal to 7 in our decimal system, or 7_{10}. For example, player 1 might say "base 3." All the players then translate the number shown into base 3. Because $12_5 = 7_{10}$, which equals $(2 \times 3^1) + (1 \times 3^0)$, 12_5 becomes 21_3.
3. Players compare their answers. Players with the correct answer get 2 points.
4. Players take turns turning over the top card from the deck and naming a base. When the card deck is gone, replace it with the shuffled discard deck.
5. The game goes on until one player has 20 points, or for as long as players choose to continue. The winner is the player with the most points.

✏ Before you play the game, try these warm-up problems.

1. Monika turns up a card that says 24_6, and names base 8. What is the answer?

   A 16_8 **B** 20_8 **C** 30_8 **D** 32_8

2. Trenton turns up the number 100_2, and names base 5. Warren gets 4_5. Benton gets 25_5. Who is correct? Explain the other player's mistake.

   Answer: ____________________ is correct.

   ____

Currency Translations Cards

Starter set of cards. Players can create additional cards. Remember that the largest digit used in a system is always one less than the base. (For example, in base 6, the largest digit you can use is 5.)

01_2	10_2	101_2	111_2	1101_2
4_5	12_5	23_5	34_5	144_5
5_6	15_6	34_6	55_6	342_6
7_8	14_8	57_8	65_8	271_8
2_{10}	24_{10}	73_{10}	96_{10}	319_{10}

Lesson 3

A Step to Your Future

Example Cecilia has a ticket for Flight 131. The round-trip ticket costs \$379. Usually 3% of the ticketed passengers for flight 131 don't show up. So the airline overbooks seats by 3%. If everyone shows up, the airline offers incentives to people who give up their seats. Cecilia gives up her seat and gets a \$200 voucher. She uses the voucher on a later trip of the same price. What percent of the cost of her later trip does she save?

Solve

Step 1: Underline the sentence that tells you the cost of her ticket.

The round-trip ticket costs \$379.

Step 2: Next, underline the words that tell you the value of the voucher.

...gets a \$200 voucher.

Step 3: Now, calculate the percent of the later trip that she saves.

$(\$200 \div \$379) \times 100\% = 52.77\%$

Answer the Question

Step 4: Cecilia saves 52.77% of the cost of her later trip.

Now try these problems.

1. Joey has a ticket for flight 57. The round-trip ticket costs \$500. The airline overbooks seats by 5%. They offer incentives to people who give up their seats when too many passengers show up. Joey volunteers to give up his seat. He gets a \$250 voucher to use in the future. What percent of a repeat trip, costing the same, does he save when he uses the voucher?

   Answer: Joey saves _______% of a repeat trip at the same price.

2. 94% of the time, 3% of the ticketed passengers for flight 245 don't show up. 98% of the time, 2% of the ticketed passengers for flight 245 don't show up. 100% of the time, 1% of the ticketed passengers for flight 245 don't show up. When full, flight 245 carries 300 passengers.

a. Based on these statistics, the probability that all of the ticketed passengers will show up is ______.

b. Make this number line to show approximately 3% of the passengers.

0 passengers ———————————————— 300 passengers

3. Sergio reads an advertisement while waiting for his luggage. It says, *"Carry a suitcase that fits in the overhead compartment or under the seat. Take 50% off your arrival time!"* It usually takes Sergio 45 minutes to get through the baggage claim area and out to his car on a Friday evening. How much time might he save by following the advice in the advertisement?

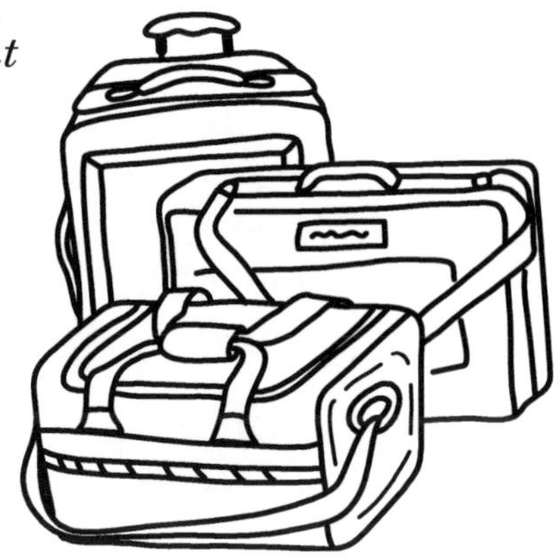

A 11.25 minutes

B 22.5 minutes

C 33.75 minutes

D 45 minutes

4. Kara goes to a 3-day business conference. Her company gives her $25 per day for meals. Each day she spends $7.50 for breakfast and $4.75 for lunch. The first two evenings she spends $14.50 and $16.80 for dinner. How much money does she has left for dinner on the third evening? Write the number of coins and bills Kara has to show the answer.

______ ______ ______ ______

______ ______ ______ ______

☆ *Challenge Problem*

You may want to talk this one over with a partner.

What is the 2004th positive odd number? Write an expression to explain your reasoning.

Lesson 4

Renting What You Need

Example Bill and his band need a sound system so they can play at a wedding. They can rent the system for one week at the location of the wedding for $185. James says they should take their own system with them. He says that the airline fee for extra baggage will only cost them $50 more each way. What is the difference in the cost? Suppose *you* have to make the decision to rent or to take the sound system. What factors might contribute to your decision?

Solve

Step 1: Underline the words that tells you the cost of renting the sound system.

They can rent the system for one week... for $185.

Step 2: Next, calculate the extra fee on the airline when they take their system.

$50 × 2 = $100 The fee is $50 more *each* way.

Step 3: Find the difference in the cost of taking their system and renting it for the week.

$185 – $100 = $85

Step 4: Think about factors that might contribute to a decision to take or rent the sound system.

Consider money, quality, ease of travel, possible damage.

Answer the Question

Step 5: The difference in the cost is $85. Factors that might contribute to my decision include these: money I have available; how important it is to have my own system; the quality of the rented system; ease of travel with carrying the system; possible damage to, or loss of, my sound system on the trip.

Now try these problems.

1. Kalani rents ski equipment for her week of skiing in the mountains. The rent is $305. If she takes her own equipment,

the baggage fee will only cost her $40 more each way. What is the difference in the cost? Why might she rent ski equipment?

Answer: The difference in the cost is $________. She might rent equipment because ____________________________________.

2. At the car rental counter, the assistant offers Riley a collision damage waiver (CDW). This promises that the company will pay for damage to a vehicle in case of an accident. Riley is in a hurry and does not read the fine print. He does not check to see if his own policy covers damage to a car he rents. He pays an extra $10 per day on the rental cost of a car. At this rate, how much does collision coverage cost per year?

   A $10 **B** $120 **C** $520 **D** $3,650

3. Beyonce goes to a regional skateboard championship. She has three options.
   i. If she pays for her hotel room three months ahead, the cost is $315.
   ii. If she pays for her hotel room two months ahead, the cost is $390.
   iii. If she pays for her hotel room at the last minute, the cost is $420.
   Complete the table. The first column tells which rates she compared to find the fraction in the second column. Then calculate the amount of money that Beyonce saves for each choice.

Comparing choices	Fraction that shows the comparison	Amount of money that Beyonce saves
Choosing option i instead of option iii	$\frac{3}{4}$	
	$\frac{21}{26}$	
	$\frac{13}{14}$	

4. Cason goes on a job interview. In the afternoon, he rents an electric cart to tour the town. The rental cost is $25 for the first hour, and $15 for each additional hour or part of an hour. He picks up the electric cart at 2:15 p.m. and returns it at 5:30 p.m.
   a. How much does he pay for his tour?
   b. Suppose he keeps the cart for another 45 minutes. How much *more* will he pay? Draw a line that connects the two correct answers.

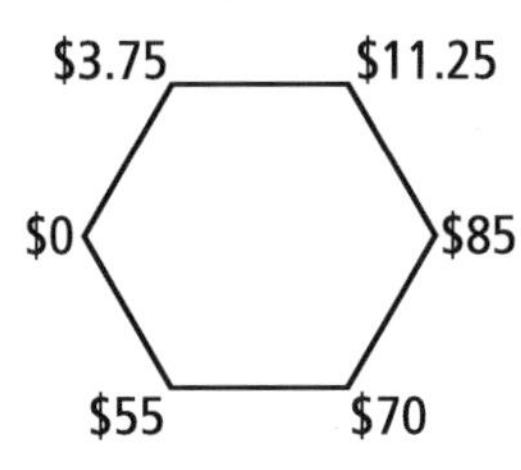

☆ *Challenge Problem*

You may want to talk this one over with a partner.

A group of students meets in the parking lot after the competition. Twelve girls leave on the first minibus. Twice as many boys as girls are left. Then 15 boys leave. Now, three times as many girls as boys remain. How many boys and girls were in the original group?

____

Lesson 5

Going Abroad

Example Penelope takes \$250 to a bank at London's Heathrow International Airport. She wants to get pounds sterling. Today, the exchange rate for dollars to pounds sterling is \$1.6272 to £1. How many pounds sterling can she get for \$250?

Solve

Step 1: Use the exchange rate to write the ratio of dollars to pounds sterling.

$$\frac{\$1.6272}{£1}$$

Step 2: Next, write a ratio for \$250 to x pounds sterling. The variable x represents the number of pounds sterling that Penelope can get for \$250.

$$\frac{\$250}{£x}$$

Step 3: Now, write a proportion to find the value of x.

$\frac{\$250}{£x} = \frac{\$1.6272}{£1}$ Set the two ratios equal.

$250(1) = x(1.6272)$ Find the cross products.

$153.64 = x$

Answer the Question

Step 4: Penelope can get £153.64 for her \$250.

Now try these problems.

1. Cedric takes £125 to a bank in New York. He wants to exchange pounds sterling for dollars. Today, the exchange rate is \$1.5912 to £1. How many dollars can he get for £125 pounds sterling?

   Answer: Cedric can get \$____________.

2. Morgan wins a place in the international surfing contest in Australia. Before leaving the country, she goes to the bank to get travelers checks. The fee for travelers checks is 3% of the amount of the purchase. The fine print says that this fee cannot be less than \$5, and will not exceed \$50. Morgan buys travelers checks

and pays a fee of exactly 3% (not more and not less). Write the range for the value of the checks she could buy.

Answer: From $____________ to $____________

3. Bonita's bank charges a fee of $1.50 each time she uses an Automatic Teller Machine (ATM) owned by another bank. While she is in Europe, she uses an ATM at a foreign bank seven times. She takes out a total of $600 in cash. What can she expect to be the effect on her account?

   A a debit of $589.50 **C** a debit of $610.50

   B a credit of $589.50 **D** a credit of $610.50

4. In one international airport, Raphael reads that bags are valued at $20 per kilogram. He remembers reading somewhere else that bags are valued at $9.07 per pound.

   a. How can he check to see if these statements mean the same thing?

   Answer: ____

   b. If a bag is not weighed in at check-in, the airline assigns it a weight of 70 pounds. Raphael loses his luggage. One bag was weighed in at 145 pounds. The other was not weighed in. Raphael writes an equation for the amount of money that he can claim. Is his equation correct? If not, make any necessary corrections.

$$(145 \times \$9.07) + (145 - 70)\$20 = \$2{,}815.15$$

☆ *Challenge Problem*

You may want to talk this one over with a partner.

The sum of three numbers is 34. The ratio of the first to the second number is $\frac{3}{4}$. The ratio of the second number to the third number is $\frac{2}{5}$. What is the second number? How did you find the answer?

____

____

____

____

____

Unit 7 Review

Review What You Learned

In this unit you have used mathematics to solve many problems. You have used mental math and estimation, practiced basic operations, solved equations, and used statistics, ratios, and proportions.

These two pages give you a chance to review the mathematics you used and check your skills.

✔ Check Your Skills

1. Ruma plans a 10-day, 9-night cruise. The cruise costs $635. The round-trip plane ticket to get to the cruise ship costs $152. She allows $500 for other items. Suppose she spreads the cost of the vacation equally over 12 months. Write an expression for the effect on her budget each month. What is that amount?

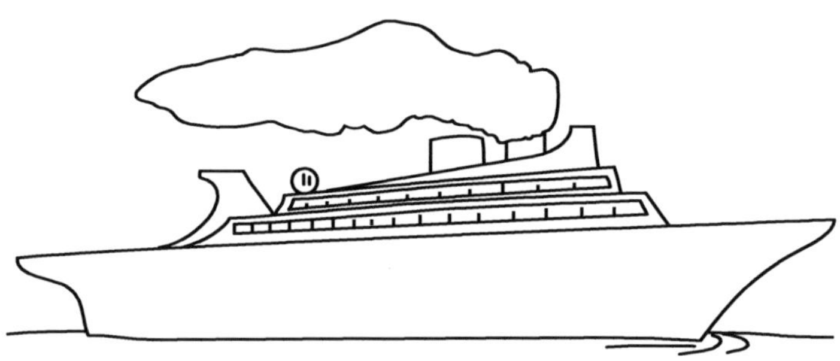

Answer: ____________________

If you need to review, return to lesson 1 (page 80).

2. Arlan has a ticket for flight 67. The round-trip ticket costs $250. The airline has overbooked the seats, and Arlan volunteers to give up his seat. He gets a $200 voucher to use on a future trip. What percent of a future trip that costs the same amount does he save? Draw a diagram to show this percent.

If you need to review, return to lesson 3 (page 84).

3. Azita rents windsurfing equipment when she goes to the beach. The rent is $106 for a weekend and $18 a day for weekdays. If she takes her own equipment, the extra baggage fees cost her $46.50 more each way. She wants the equipment for Saturday, Sunday, and Monday. Azita takes her own equipment. What fraction of the rental cost does she save? Shade an area of the circle that represents this fraction. Show your calculations.

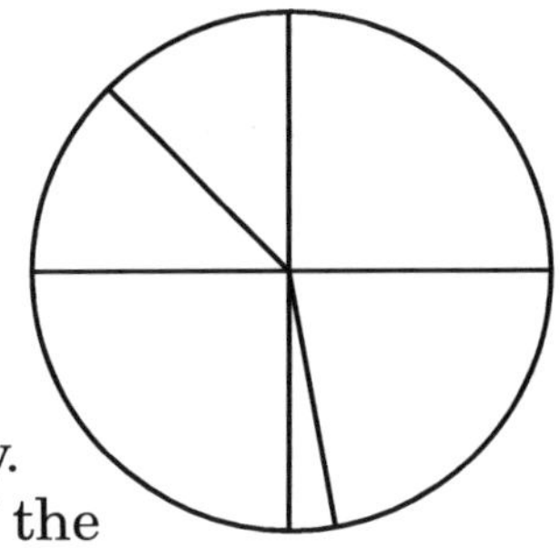

Answer: ______________________________

If you need to review, return to lesson 4 (page 86).

4. Today, the exchange rate for dollars to English pounds sterling is $1.6272 to £1. How many pounds sterling can you get for $500?

   A £32.50   **C** £500

   B £307.28   **D** £813.60

If you need to review, return to lesson 5 (page 88).

5. Garrett reads at the airport that bags are valued at $9.07 per pound. If a bag is not weighed in at check-in, the airline assigns it a weight of 70 pounds. Garrett loses his carry-on bag. He thinks it weighs 100 pounds. It was not weighed in at check-in. Write an equation for the amount of money he can claim.

Answer: ______________________________

If you need to review, return to lesson 5 (page 88).

Write Your Own Problem

Choose a problem you liked from this unit. Write a similar problem using a situation and related facts from your own life. With a partner, share and solve these problems together. Discuss the mathematics and compare the steps you used. If you need to, rewrite or correct the problems. Write your edited problem and the answer here.

____

____

____

____

Unit 8 — Keeping the Books

Preview

How You Will Use This Unit

Keeping the books, or financial records, involves many different things. You may track income and expense. You may need to decide what software to use. You may also think about how to keep the books for a club or organization, and even for a large event. As you compare options and make choices, you will often use math. The math skills you use include mental math and estimation, basic operations and equations, statistics, and ratios and proportions.

What You Will Do in This Unit

In this unit, math steps demonstrate how to solve problems. These steps can help you answer questions such as these:

You receive $1,714 in salary. You get paid for an invoice of $300 for this month's tutoring. You pay bills that add to $1,092. You are also paid for new invoices that add to $986. Extra expenses total $623. What are your income and expenses?

Your club decides to go on a bike trip. You estimate the cost of various items for the trip. After the trip, you compare each estimate to the actual cost for each item. What is the variation for each item?

You review the financial results of the summer festival. You make a table of major sources of income and the major categories of expense. After that, can you see if the event made or lost money?

You use a spreadsheet software package to manage your finances. What are some of the advantages that you are looking for?

What You Can Learn from This Unit

When you complete this unit, you will have used mathematics to work problems related to keeping the books. These problems are similar to those that may actually occur in your daily life.

Lesson 1 Keeping Records

Example During January, Mariano receives $1,404 as his salary. He is also paid $543 rent on the upstairs apartment. He pays his mortgage, utility, telephone, and credit card bills for a total of $1,012. He receives payment for this month's invoices for a total of $2,314. Miscellaneous expenses add to $655. Answer these questions for this month.

a. What is his total income?

b. What are his total expenses?

c. What is the difference between his income and expenses?

Solve

Step 1: List all Mariano's income items and find their total.
Income items → $1,404 + $543 + $2,314 = $4,261

Step 2: Now, list Mariano's expenses and find that total.
Expenses → $1,012 + $655 = $1,667

Step 3: Find the difference.
$4,261 – $1,667 = $2,594

Answer the Question

Step 4: **a.** His total income is $4,261.
b. His total expenses are $1,667.
c. His income is greater than his expenses by $2,594.

✏ Now try these problems.

1. During September, Laurel receives $1,521 in salary. She is paid $400 for her graphic arts project and $310 for an invoice. She pays her rent, utility, telephone, and credit card bills for a total of $1,143. She receives payment for additional invoices totalling $2,877. Miscellaneous expenses add to $474.

   a. What is her income?

   b. What are her expenses?

   c. What is the difference?

   Answer: **a.** Laurel's income is $__________.

   b. Laurel's expenses are $__________.

   c. The difference is that her __________ is greater than her __________ by $__________.

2. This month Patience receives a credit card bill for $1,906. She pays $25 on this bill. What is the difference between the amount she pays and the amount she owes?

Answer: $___________

3. Elyssa charges one customer $15 per hour for her time. She keeps a record of her hours. Her client offers to pay her a fixed fee of $1,200 for a similar project in the future. Suppose you are Elyssa. Would you accept this offer? Why? Why not?

March	Monday	Tuesday	Wednesday	Thursday	Friday	Saturday
Week 1—hours:	0.0	0.0	8.0	7.5	9.0	9.0
Week 2—hours:	5.0	6.5	8.0	8.0	7.5	7.5
Week 3—hours:	7.5	10.0	9.0	0.0	0.0	0.0

Answer: ___________________________

4. Cody transfers the balances from his three credit cards to a new *Dynamo* credit card. The rate on each card is 18.9%, 12.45% and 14.67%. The new card company offers a rate of 0% interest on all transactions for 12 months. What are some advantages and disadvantages of transferring his balances to this new card?

Advantages: ___________________________

Disadvantages: ___________________________

☆ *Challenge Problem*

You may want to talk this one over with a partner.

You have $7.85 worth of nickels and dimes in the petty cash drawer in your business. You have an odd number of nickels and more dimes than nickels. What is the largest number of nickels that you can have? Explain how you get your answer.

____

____

____

____

____

____

Lesson 2 — Club Records

Example *ClubAdventure* decides to go on a field trip to the marine science center. Camille estimates the cost of various items for this trip. The club raises money to pay for the trip. After the trip, Camille compares the estimated cost to the actual cost for each item. What is the difference for each item? Did Camille underestimate or overestimate the total cost of the trip?

Field Trip Finances	Estimated cost	Actual cost	Difference (use a minus sign when actual > estimate)
Hotel	$1,350	$1,390	
Transportation	$255	$250	
Entrance tickets	$240	$240	
Food	$400	$382	

Solve

Step 1: Subtract the actual cost from the estimated cost for the hotel.
$1,350 – $1,390 = –$40

Step 2: Subtract the actual cost from the estimated cost for the other items.

Transportation $255 – $250 = $5
Entrance tickets $240 – $240 = $0
Food $400 – $382 = $18

Step 3: Enter these values in the table.

Difference
–$40
$5
$0
$18

Step 4: Calculate the total variation.
–$40 + $5 + $18 = –$17

Answer the Question

Step 5: Camille underestimated the cost of the trip by $17.

Now try these problems.

1. *ClubSport* decides to go on a ski trip. Michael estimates the cost of various items. The club raises money to pay for the trip. After the trip, Michael compares the estimated cost to the actual cost

for each item. What is the difference per item? Did Michael underestimate or overestimate the cost of the trip?

Ski Trip Finances	Estimated cost	Actual cost	Variation
Hotel	$2,160	$2,135	
Transportation	$2,413	$2,567	
Ski tickets	$2,100	$2,400	
Food	$850	$545	

Answer: Michael __________ (underestimated/overestimated)

by $__________.

2. The lunch special at the *OnTheBeach* cafe costs $4.95. The *Theta4* club has $53.00 in its club account. What is the greatest number of lunch specials that the club can buy with this money?

   A 8 **B** 9 **C** 10 **D** 11

3. The student council budgets $840 for beach cleanup events this year. Last year, the records showed a total of $700 spent on these events. Which sector of this circle graph shows the percent increase from last year to this? Shade and label that sector.

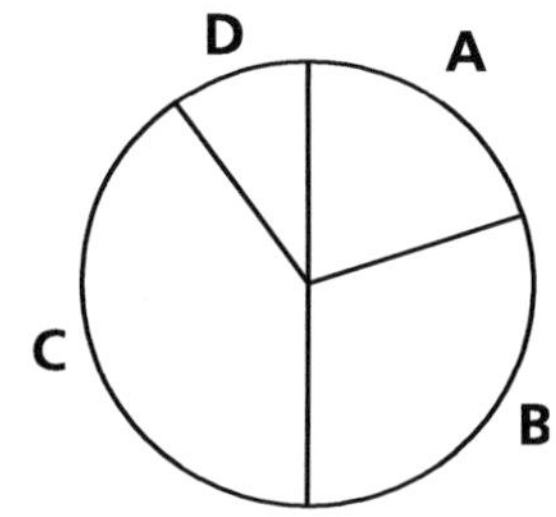

4. The art center's postage cost went from $238.00 in January to $117.60 in February. Complete the equation and find the percent decrease in postage.

   Answer: ($________ – $________) ÷ $________ × 100% = _______%

☆ *Challenge Problem*

You may want to talk this one over with a partner.

Fill in the empty squares with one-digit numbers (choosing from 0 through 9). The numbers in the unshaded squares of each row or column must multiply to give the number in the shaded square that begins the row or column.

	24	15	2
9			
10			
8			

Lesson 3

Bingo Books

A Calculation Game (for Two or More Players)

The goal of this game is to be the first to cover a complete line of answers with your counters. First, someone gives a math problem. Then, you find the answer on your *Bingo Books* sheet.

Materials

Bingo Books sheets (next page); counters; paper and pencils; board for writing the problem

Directions

1. Each player has a *Bingo Books* sheet, a handful of counters, a pencil, and scratch paper. Players sit where they can see the board.
2. Player 1 writes a math problem on the board. The problem must have an answer that is a number from 1 to 60. For example, $9 \times 8 \div 18$. Other players first calculate the answer (for example, 4), and then find the answer on their *Bingo Books* sheets, and place a counter on that square. Player 1 then tags another student to write the next math problem.
3. Players take turns writing math problems as they are tagged. Each one must have a whole number answer from 1 to 60. A player can refer to the answer from the previous problem. For example, "What is the square root of the last answer?" The other players find the answer (for example, 2) and cover that number on their *Bingo Books* sheets.
4. The winner is the first player to get a complete vertical or horizontal line of correct answers. Players check the answers and start again with new *Bingo Books* sheets.

Before you play the game, try these warm-up problems.

1. Anson writes the problem, $4^2 \times 1\frac{1}{2}$. Warren covers the number 8. Lance covers the number 24. Who covers the correct number?

   Answer: ________________

2. Anson tags Taylor to write the next problem. Taylor writes this problem: "What number is a factor of the last answer and is also 3 greater than another factor of that answer?"

   A 3 **B** 4 **C** 6 **D** 12

Bingo Books Sheets

Bingo Books sheets can have many different patterns of numbers, but always have 1 through 60. Each sheet must have the same dimensions (columns by rows). Here are two examples.

Players can have identical or different sheets.

1	2	3	4	5	6	7	8	9	10
11	12	13	14	15	16	17	18	19	20
21	22	23	24	25	26	27	28	29	30
31	32	33	34	35	36	37	38	39	40
41	42	43	44	45	46	47	48	49	50
51	52	53	54	55	56	57	58	59	60

16	27	28	39	40	50	51	57	58	60
15	17	26	29	38	41	49	52	56	59
7	14	18	25	30	37	42	48	53	55
6	8	13	19	24	31	36	43	47	54
2	5	9	12	20	23	32	35	44	46
1	3	4	10	11	21	22	33	34	45

Lesson 4 Big Events

Example Vrae reviews the results of the costs and income for the *Summerfest.* He makes a table of major sources of income and major categories of expense. What are the total income and total expenses? Does the event make or lose money?

Income		
Item	**Tickets Sold**	**Total Income**
Participant fee, \$25	32	
Vendor fee, \$15	50	
Entertainment:		
TGIF band, \$5	75	
Saturday, *Brownbaggers* lunch band, \$8	150	
Saturday, *NiteWild* band, \$12	210	
TOTAL INCOME		
Expenses		**Total Expenses**
Set up, tear-down, cleanup		\$200
TGIF band		\$250
Brownbaggers band		\$400
NiteWild band		\$850
TOTAL EXPENSES		

Solve

Step 1: Find the income for each item. Then, find the total income.

\$25 × 32 = \$800 — 32 participants who pay a \$25 fee each.

\$15 × 50 = \$750 — 50 vendors who pay a \$15 fee each.

\$5 × 75 + \$8 × 150 + \$12 × 210 = \$4,095 — Income from bands

TOTAL: \$800 + \$750 + \$4,095 = \$5,645

Step 2: Now, compute total expenses.

\$200 + \$250 + \$400 + \$850 = \$1,700

Answer the Question

Step 3: Total income is \$5,645. Total expenses are \$1,700. The event makes money.

Now try these problems.

Refer to the information in the example as you work these problems.

1. Vrae reviews his income and expense data from *Summerfest.* He wonders what would happen if the weather were bad. He estimates that only one third of the people would buy tickets to listen to the bands. Suppose all the other data stays the same.

What would the total income and expenses be? Would the event make or lose money?

Answer: Income would be $________. Expenses would be $________.

The event would ________ (make/lose) money.

2. Drew manages *Summerfest* the following year. He reduces the participant fee by 10% and increases the vendor fee by 5%. Suppose that the same number of participants and vendors take part.
   a. Write an expression to show the difference in income from the previous year.
   b. How much more or less will the event make than the previous year?

   Answer: a. ____________________ **b.** $__________

3. Drew retains two bands who charge $350 and $900 to play. He predicts that 80 people will pay $12 to listen to one band, and 150 people will pay $16 to listen to the other band. What would the income and expenses for the entertainment be using his predictions?
   - **A** Income: $3,360; expenses: $1,250
   - **B** Income: $3,360; expenses: $550
   - **C** Income: $1,250; expenses: $3,360
   - **D** Income: $3,080; expenses: $1,250

4. Jayfus advertises early registration for *Summerfest* the following year. He prepares this poster which offers advance tickets to *both* of Drew's concerts at 25% off. 75 people buy these advance tickets. On the day of the event, 60 people buy $12 tickets, and 150 people buy $16 tickets.
   a. What is the price for a set of advance tickets?
   b. How much income does the entertainment part of the event actually make?

   Answer: a. $__________ **b.** $__________

Single tickets $12 and $16 Attend both concerts and save 25%!!

☆ *Challenge Problem*

You may want to talk this one over with a partner.

You and some friends organize a big event. You manage the bands. One friend does the tickets. A second friend sees to the participants. A third friend manages vendors. A fourth friend organizes setting up and cleaning, and tracks the income and expenses for the event. What are advantages and disadvantages of each person managing a different part of the event?

____

____

____

Lesson 5

Spreadsheets & Software

Example Jonelle uses a spreadsheet to keep track of finances at her daycare center. Each month she enters the amount of money that she collects for each child. The spreadsheet software gives her the total per month. It also gives the total per child since January. Calculate the total for each child during the quarter (the first three months of the year). What does the TOTAL line of the spreadsheet look like at the end of the first quarter? Does the *1st quarter total* for the five students match the total for the three months in the last row?

Child's name	January	February	March	1st quarter total
Alexandra	\$350	\$350	\$350	
Chase	\$275	\$300	\$225	
Logan	\$380	\$275	\$275	
Michael	\$125	\$200	\$250	
Stacia	\$420	\$325	\$325	
Total				

Solve

Step 1: ***Columns*** Use mental math to calculate each monthly totals. Then find the three-month sum. Notice the amounts are grouped for easy adding.

January → \$350 + (\$275 + \$125) + (\$380 + \$420) = \$1,550

February → \$350 + (\$300 + \$200) + (\$275 + \$325) = \$1,450

March → (\$350 + \$250) + (\$225 + \$275) + \$325 = \$1,425

Total for the quarter → \$1,550 + \$1,450 + \$1,425 = \$4,425

Step 2: ***Rows*** Use mental math to calculate the total collected for each child. Then find the sum for all five children.

Alexandra → \$350 × 3 = \$1,050

Chase → (\$275 + \$225) + \$300 = \$800

Logan → \$380 + \$275 × 2 = \$930

Michael → \$125 + \$200 + \$250 = \$575

Stacia → \$420 + \$325 × 2 = \$1,070

Total for the quarter → \$1,050 + \$800 + \$930 + \$575 + \$1,070 = \$4,425

Answer the Question

Step 3:

Total	\$1,550	\$1,450	\$1,425	\$4,425

The column total *does* match the row total.

Now try these problems.

Refer to the information in the example as you work the first problem.

1. During March, Jonelle adds a sixth child and collects a $200 payment.
   a. What does the TOTAL line of the spreadsheet now look like?
   b. What is the new sum for the six totals in the *1st quarter total* column? Does it match the row total for the three months? Why?

Answer: a.

Total				

b. The new sum of the *1st quarter total* column is $________.

It __________ (does/does not) match the row total because

____.

2. Bryant compares the price of several software packages. He calculates the mean price, and buys the package that is priced closest to this number. Which package does he buy?

Fingertip Finances	***Fast Finances***	***Finances First***	***Financial Views***
One package does it all	*BEST SELLER*	*Everything you need*	*Tailor-made to your needs*
Only $209	**$299**	**$599**	**$850**

A *Fingertip Finances* **C** *Finances First*

B *Fast Finances* **D** *Financial Views*

3. Lucianne conducts a survey to find out what activities are most popular. She uses her software package to present the information in a circle diagram. Out of 425 responses, about how many responses name each activity?

Answer: ____ choose basketball. ____ choose softball.

____ choose robotics. ____ choose chess.

____ choose debate. ____ choose band.

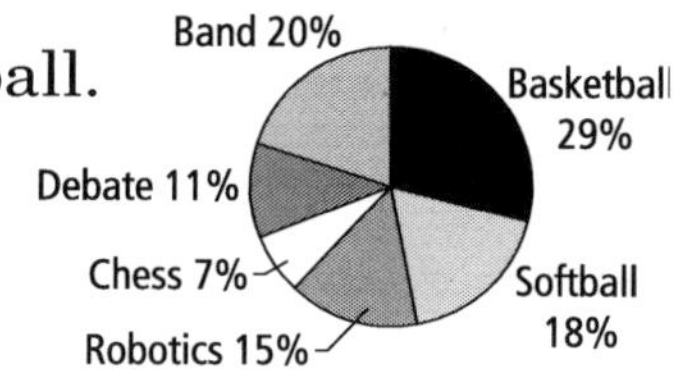

☆ *Challenge Problem*

You may want to talk this one over with a partner.

You use a software package to help gather, manage, and present data from a community project. The package allows you to look at the data as tables, graphs, and charts. You can calculate percents and relationships among the data with the software. What are some advantages and disadvantages of using software for this sort of project?

____

____

____

Unit 8 Review

Review What You Learned

In this unit you have used mathematics to solve many problems. You have used mental math and estimation, practiced basic operations, solved equations, and used statistics, ratios, and proportions.

These two pages give you a chance to review the mathematics you used and check your skills.

✔ Check Your Skills

1. During November, Jeannianne receives $1,714 in salary and $250 for tutoring after school. She receives payment for an invoice of $165 for tutoring at home. She pays her rent, utility, telephone, and credit card bills for a total of $1,092. She also gets paid for other invoices that total $986. Miscellaneous expenses add to $623.

   a. What is her income?

   b. What are her expenses?

   Answer: Income → ______________________

   Expenses → ______________________

   If you need to review, return to lesson 1 (page 93).

2. *ClubArt* decides to go on a photography trip to the botanical gardens. Daran estimates the cost of various items. The club raises money to pay for the trip. After the trip, he makes a table to compare the estimated cost to the actual cost for each item. Complete the table to show the difference for each item. Did Daran underestimate or overestimate?

	Estimated cost	Actual cost	Variance
Hotel	$1,675	$1,654	
Transportation	$125	$134	
Entrance tickets	$160	$140	
Food	$360	$382	

Answer: Daran _________ (underestimated/overestimated).

If you need to review, return to lesson 2 (page 95).

3. The heritage club's advertising costs went from $358 in March to $242 in April.

a. Complete the equation to give the percent decrease in advertising costs.

Answer: $________ ÷ $________ × 100% = ______%

b. Highlight the portion of the line to show approximately this percent decrease.

0% 50% 100%

4. Faith makes a table to review the financial results of the town's *FallBackfest.* Complete the table to show the total income and total expenses. Does the event make money?

Income		
Item	**Tickets Sold**	**Total**
Participant fee, $30	45	
Vendor fee, $15	25	
Entertainment:		
Friday *Bandstand,* $6	60	
Saturday's Child lunch band, $7	90	
Saturday *Freedom* band, $15	180	
TOTAL INCOME		

Expenses	
Set up, teardown, cleanup	$250
Bandstand	$300
Saturday's Child lunch band	$520
Freedom band	$650
TOTAL EXPENSES	

Answer: The event __________ (does/does not) make money.

If you need to review, return to lesson 4 (page 99).

5. Keyton uses his software package to track tickets sold for various events. Two different events are scheduled for July through September. So far he has sold tickets that total $1,365 for one event and $785 for the other event. The sum of the totals for the three months should be $________.

If you need to review, return to lesson 5 (page 101).

Write Your Own Problem

Choose a problem you liked from this unit. Write a similar problem using a situation and related facts from your own life. With a partner, share and solve these problems together. Discuss the mathematics and compare the steps you used. If you need to, rewrite or correct the problems. Write your edited problem and the answer here.

To the Teacher

Welcome to *Budgeting & Banking,* Book 4 of the *21st Century Lifeskills Mathematics* series.

Mastery of practical math skills is the overarching goal of the *21st Century Lifeskills Mathematics* series. To this end, each of the six books has been carefully designed to present topics students are likely to encounter in everyday life. Each book includes problems that involve estimation, equations, mental math, calculators, and critical thinking. Each book includes additional concept-specific skills such as graphing, averages, statistics, ratios, and measurement.

The books are appropriate for use with small groups, a full class, or by independent learners. The self-explanatory nature of the lessons frees the teacher for individual instruction. Each unit begins with a preview lesson, which models and explains the types of problems students will encounter in the unit. Then there are five lessons, at least one of which is a game. Game titles are italicized in the Table of Contents, on the lesson pages, and in the Answer Key. Each unit ends with a review of the unit concepts. Both illustrations and graphic art are used to support the instruction and maintain interest. A variety of problem types and games are used to sharpen critical thinking skills throughout the program.

Below are the titles of the other books in the *21st Century Lifeskills Mathematics* series:

Book 1: Everyday Life
Book 2: Home & School
Book 3: On the Job
Book 5: Smart Shopping
Book 6: Sports, Hobbies, & Recreation

Students from middle school through adult classes will appreciate the practical content of each book.

Through modeling, practice, and review, students will build their math skills and learn to approach everyday mathematical situations with confidence. *21st Century Lifeskills Mathematics* will help your students become successful problem solvers!

Budgeting & Banking — Answer Key

Unit 1: Paying Your Way

Lesson 1: Sharing Day-to-Day Costs

1. **a.** $\frac{6}{10}$ or $\frac{3}{5}$; **b.** $457.50
2. Show bills that total $152.
3. C $400
4. $14.00

Challenge Problem. Sample answer: Yes. $(4 \times 4 + 4) \div 4 - 4 = 1$ or $(x \times x + x) \div x - x$ because this always equals 1.

Lesson 2: Budgeting for Special Items

1. B $\frac{x}{6} = \frac{60}{1}$
2. Spread 36 hours (for 9 remaining drawings) between Wednesday and Saturday.
3. $20.05
4. Sample: ($3,600 + $1,600 x 2) ÷ (12 months × 2) = $283.33 per month

Challenge Problem. Explanations will differ. To explain, you could try using x for the original number and using algebra.

Lesson 3: Saving, a Little at a Time

1. **a.** 25; **b.** 2,500; **c.** 500
2. B $10
3. Circle $600, $300, $250; Draw a square around $400.
4. Year 1: Year's Savings $2,500; Accumulated Savings: $2,500; Year 2: Income $27,000; Year's Savings $2,700; Accumulated Savings: $5,200; Year 3: Income 1.08 × $27,000 = $29,160; Year's Savings $2,916; Accumulated Savings: $8,116

Challenge problem. 297

Lesson 4: Using Bank Services

1. C $f = (\$0.05 \times t) + \$4$
2. Answers may vary. Key facts include the number of monthly transactions and that Jere's bank will charge her a fee of $5.80 this month. The bank across the street would only charge $5.00.
3. Yes $1,483.82 – $525.00 – $745.00 = $213.82
4. Circle $1.

Challenge Problem. Answers will vary. Some key facts include how you use the debit card; whether you need more than $500 in the account ($2,000–$5,000); whether you like to pay off your credit card debts without incurring interest charges; current versus future expenses and income.

Lesson 5: *It All Adds Up!*

1. A 1
2. 4; the corner squares each have 2 open edges.

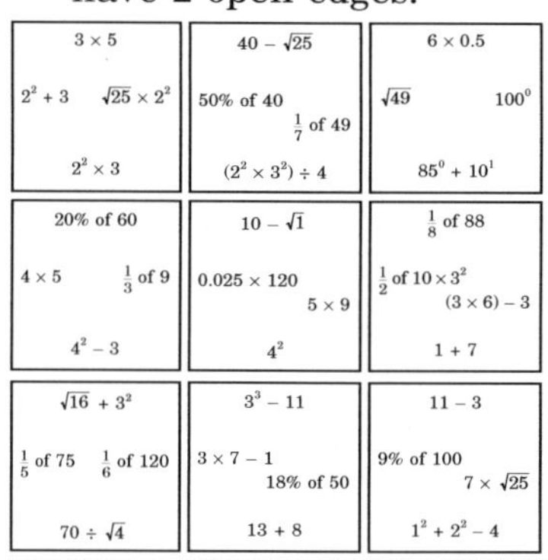

Review

1. **a.** $\frac{11}{20}$; **b.** $570
2. B $\frac{x}{3} = \frac{125}{1}$
3. **a.** Circle 100% inside the circle;
   b. Circle $10,000 outside the circle.
4. D $f = \$0.05 \times t + \$5$

Unit 2: Buying Large Items

Lesson 1: Buying a Car

1. $3,000
2.

0%-interest plans	Pay off in 1 year	Pay off in 2 years	Pay off in 3 years
$1,000 down payment	$1,325/ month	$662.50/ month	$441.67/ month
$2,000 down payment	$1,241.67/ month	$620.83/ month	$413.89/ month
$5,000 down payment	$1,158.33/ month	$579.17/ month	$386.11/ month

Faye computed these monthly payments based on a $3,000 down payment.

3. B $12,740
4. Car 1 Insurance: $1,600; Total cost: $7,000.
   Car 2 Car cost: $6,600; Insurance: $1,456; Total cost: $8,056.
   Answers will vary. Key facts include value of reduced insurance cost, value of car for resale, value of additional accessories, amount of money that you can afford.

Challenge Problem. $54.45

Lesson 2: Buying a Home

1. $210
2. C $450
3. 28%; $560; $560; $80,000
4.

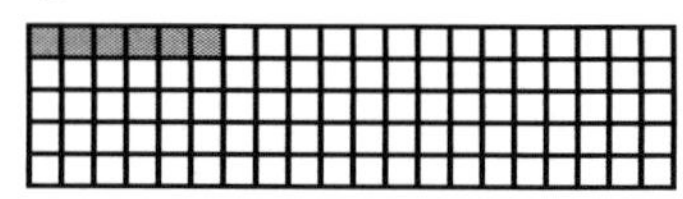

Challenge Problem.

Taxing Body	Tax rate	Taxes owed
State	0.02%	$13
County	0.62%	$403
Town	0.44%	$286
School District	1.47%	$955.50
Library	0.12%	$78
TOTAL ANNUAL TAXES		$1,735.50

Lesson 3: College Costs

1. D $4,708.90
2. Amount owed after 2 years = $3,300

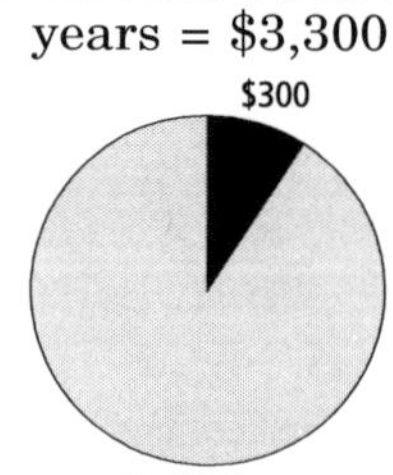

3. ($2,000 × 1.02) + ($1,500 × 1.05) = $3,615
4. D $4,500 + 6(1,500 × 0.05)

Challenge Problem. Answer will vary. One key fact is that costs typically increase with the age of the child.

Unit 2 *(continued)*

Lesson 4: *What Do You Owe?*

1. 9
2. C 5

Lesson 5: Buying Insurance

1. a. $\frac{1}{9}$; b. $160
2. B $3,300
3. a. $288
   b. $160
   c. not buying
4.

Years after purchase	Time left on guarantee / total years	Value
1	$\frac{4}{5}$	$228 (circled)
2	$\frac{3}{5}$	$171
3	$\frac{2}{5}$	$114
4	$\frac{1}{5}$	$57
5	0	$0

Challenge Problem. Answers will vary. Based on last year, key facts include cost of towing without towing coverage ($160) versus cost of towing with towing coverage ($53) and the probability of requiring a tow with an aging vehicle.

Review

1. $x \times 24 = p$; $\$2{,}500 + (x - \$50) \times 20 = p$; $375; $9,000
2. C $0.007 \times \$40{,}000$
3.

Year	Balance at start of year	Interest incurred	Balance at end of year
1	$6,000	$390	$6,390
2	$6,390	~~$830.70~~ $415.35	~~$7,220.70~~ $6,805.35

4. a. Check that diagram shows Hal's car insurance is $\frac{3}{25}$ of his monthly income;
   b. $165
5. buying

Unit 3: Unplanned Expenses

Lesson 1: Telephone & Other Bills

1. $1,320
2. a. Circle 22;
   b. Draw a square around $5.20.
3. C $80; Answers will vary. One possible answer would be to take the average (mean) of the last three months' unexpected expenses.
4.

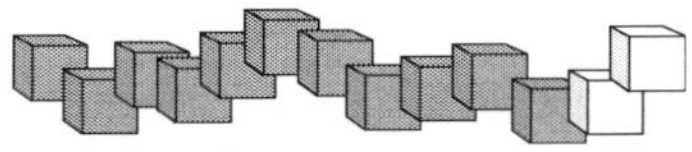

Challenge Problem. Your friend has 3. If your friend had 14, then your friend would know that you have 7, but your friend does not know, so your friend must have 3.

Lesson 2: Doctors, Dentists, & Veterinarians

1. $6
2. B 10
3. ($110 + $650 + $35) ÷ 6 = $132.50
4.

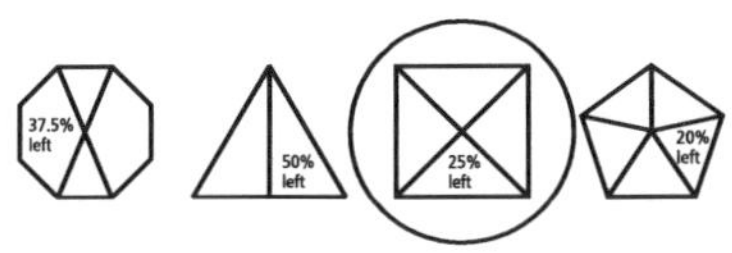

Challenge Problem.

```
        1               1
       3 5              4
     7 9 11             9
   13 15 17 19         16
 21 23 25 27 29        25
```

10,000. Answer is the square of the row number.

Lesson 3: *Careful Spending*

1. 1, 1
2. A 5 in row 1, column 5

2	1	2	~~2~~	5	~~4~~
5	1	1	~~6~~	~~2~~	3
~~3~~	~~3~~	5	1	3	1
1	7	~~6~~	1	1	~~2~~
2	~~5~~	~~1~~	4	1	3
~~4~~	1	2	4	~~2~~	3

Lesson 4: Unplanned Repairs

1. $x/(\$180 + x) = \frac{1}{5}$
2. Answers will vary. One answer is that she saves $130 by not having the bug shield catch replaced. Another is that she saves $130 plus the $20 she made. On the other hand, she no longer has the shield.
3. A $117.50
4.

Month		0	1	2	3	4	
Balance of repair bill	$520	$420	$370	$320	$270	$220	
Emergency fund	$350	$250	$225	$200	$175	$150	
Month	5	6	7	8	9	10	11
Balance of repair bill	$170	$120	$70	$20	$0	$0	$0
Emergency fund	$125	$100	$75	$50	$55	$80	$105

Challenge Problem. Answers will vary. Most would agree that they would pay the $20 charge plus a donation, perhaps another $20, and more, if possible, to support the school.

Lesson 5: Gifts, Flowers, & Unexpected Trips

1. $\frac{1}{12}$;

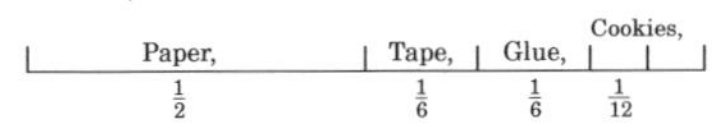

2. C $20.56
3. 3.75 more
4. ($18 × 10 + $72.50 + $59.50) ÷ 12 = $26

Challenge Problem. The advertised price is $\frac{2}{3}$ the normal price. Overhead (building, sales clerks) may be less.

Review

1. $182
2. $0; $8.99 ($8.99 + $0); $8.99
3. B $150
4. Delfino has $\frac{1}{4}$ of his original money left afterwards. Diagram should show that what he has left is $\frac{1}{4}$ of the whole.

Unit 4: Credit Card Budgeting

Lesson 1: How Credit Cards Work

1. $206.24
2. $342.24

Month	1	2	3	4	5	6	7	8	9
Payment	$50	$50	$50	$50	$50	$50	$42.24	$0	$0
Amount owed	$279.95	$234.09	$187.40	$139.87	$91.49	$42.24	$0	$0	$0

3. A 0.015 × 12 × $1,000 = $180

Unit 4 *(continued)*

4. (.014 × \$1,500 × 12) + \$40 = \$292

Challenge Problem. Below \$777.78 because (12.5% of x) must be less than [(8% of x) + \$35]

Lesson 2: *Percents*

1. Either $\frac{1}{4}$ (you complete the set and win 5 points) or 30% to open another set.
2. C 62.5%

Lesson 3: Paying by Credit Card

1. a. Yes, saves
   b. About \$173.33 (depending on how you round)
2. a. Unpaid balances:
   1: \$90; 2: \$181.35;
   3: \$274.07; 4: \$368.18;
   5: \$463.70; 6: \$560.66;
   7: \$659.07; 8: \$758.96;
   9: \$860.34; 10: \$963.24;
   11: \$1,067.69;
   12: \$1,173.70
   b. Answers will vary. The key fact is that his debt increases unless he increases payments.
3. D *MyCharge;* payment due in 50 days

Challenge Problem. Answers will vary. Key factors include the difference in the actual cost of reservations (\$290.78 by credit card versus \$325 with cash); ensuring you have a reservation; building a credit rating.

Lesson 4: Credit Card Services

1. \$160
2. \$12,000/year. Key points include the interest that Felicia pays on her unpaid balance (which increases what she really pays for the ticket) and the cost of buying a regular ticket. Compare this to what she pays for the ticket through her credit card.
3. B \$87.60
4. a. Draw a circle around \$25
   b. Draw a square around \$600.

Challenge Problem. \$200

Lesson 5: Reading the Fine Print!

1.

Original Unpaid Balance	Late Payment Fee	Interest Owed	Payment	New Unpaid Balance
\$155	\$25	\$2.79	\$10	\$172.79

2. Circle \$54 (for the month in which he goes over the limit and the following month when he is late).
3.

Balance	\$1,239.00	\$1,770.00	\$92.12
Transaction Fee	\$37.17	\$50	\$5

4. C \$309

Challenge Problem. Answers will vary. Key factors include whether you minimize the interest you pay on any unpaid balance. Also whether you will use the services and how much each service will cost (when you include the annual fee and any other hidden costs of use).

Review

1. (\$245 – \$25) × 1.019 = \$224.18
2. Unpaid balances:
   1: \$177.83; 2: \$156.19;
   3: \$134.08; 4: \$111.48;
   5: \$88.38; 6: \$64.77;
   7: \$40.64; 8: \$15.98; 9: \$0
   a. 9 months
   b. saves; \$29.02
3. B *Thistle;* payment due in 48 days
4. \$20,000/\$2,500 × \$65 = \$520
5.

Payment	Original Unpaid Balance	Payment received by closing date	Late Payment Fee	Interest Owed	New Balance after payment
On-time	\$247.50	Yes	\$0	\$3.10	\$150.60
Late	\$247.50	No	\$35	\$5.20	\$187.70

Unit 5: Rental Budgeting

Lesson 1: Houses & Apartments

1. B \$27,279.00; \$568.31
2. \$7,200; \$3,600; \$10,800; \$13,740; does not; \$2,940
3. a. Diagram should show a rectangular shape with dimensions such as 5 feet by 10 feet
   b. \$1.50

Challenge Problem.

	3	4	5
	9	6	3
1	3	0	8

F = 4 or 5; N = 9; S = 6; O = 1; Z = 0; E = 8: The letter is the first letter of the word name for the number (as in, Four, Five, Nine, Six . . .).

Lesson 2: Renting Wheels

1. 27.5 = x + 50y and 40 = x + 100y
2. 24 Friday 2:59 p.m. Drop off minibus.
3. D Saturday, Sunday
4. About \$10

Challenge Problem. Answers will vary. Some key points might include the relative rental costs, the gas mileage, the ability of the two families to get along together in tight quarters.

Lesson 3: Tools

1. 12
2. C \$18.75
3. a.

Number of times per month that Hilde rents out the tent	Number of months before she starts making money
1	32.5
2	16.25
3	10.83
4	8.125

   b. Circle 3 times per month.
4. Answers will vary. Key factors include his costs of \$205 and the deposit of \$285.

Challenge Problem. 27

# of painted surfaces on a small cube	0	1	2	3	4	5	6
# of cubes	1	6	12	8	0	0	0

Lesson 4: *Chain of Rentals*

1. 5 hours — \$4 to match \$20; or 4 weeks — \$2,048 to match \$512

Lesson 5: Renting to Buy

1. (\$405 × 24) + (\$452.59 × 36) – \$20,499 = \$5,514.24
2. \$19,620; \$22,298.88; Answers will vary. Key points include availability

of money, relative cost of the two plans, paying sooner than later, ability to turn the car in at the end of the lease.

3. C About $3\frac{1}{2}$ years
4. **a.** 500
   b. Diagram should show a steady increase to 9,500 stores after 5 more years.

Challenge Problem. Sample answer: 8, 7, 6, 5, 4, 3, 2, 1, 0, –1, –2, –3, etc. The value of (9 × ninth term) in this sample answer is 0.

Review

1. **a.** $625; $687.50; $756.25
   b. $24,825.00; c. $689.58
2. $0.10; $35
3. 7; Find the next greatest whole number greater than $\frac{\$99}{\$15}$.
4. A $6,201

Unit 6: Self-Employment

Lesson 1: Starting Out

1. 1,350
2. $45 × $\frac{200}{50}$ or $180; $79.50 + ($12.99 × $\frac{2}{5}$) or $84.70; $95.30; Some possible assumptions are that she will be able to use the remaining ream of paper and the rest of the ink in the cartridges. There might be issues about the quality.
3. B $187.50
4. 12.5

Challenge Problem. About $318.50. Explanations should show 6% spread over 12 months ($\frac{1}{2}$% per month), calculated each month on the monthly balance for 12 months.

Lesson 2: Income versus Expenses

1. **a.** $0.20
   b.

Packet size	Sale price	Cost to make mix	Profit
8 ounces	$3	$2	$1
12 ounces	$4.50	$2.20	$2.30
16 ounces	$6	$2.40	$3.60
20 ounces	$7.50	$2.60	$4.90

   c. Non-food expenses stay constant. A packet that is 50% larger uses 50% more ingredients.
2. C probability; The probability of a customer picking one of Saharah's stands is $\frac{2}{4}$ or $\frac{1}{2}$. The probability of that customers picking one of the others is $\frac{1}{4}$.
3. About 25,755

Challenge Problem. 301. One way is to make a chart and look for a pattern. Sample expression: (n × 2) + (n + 1)

Lesson 3: Insurance & Taxes

1. A $540
2. $215 × 12 × 0.153 = $394.74
3. **a.** $50; **b.** $3,050
4. Circle center column ($2,025, $675, $823.50).

Challenge Problem. 15. Reflect the figure over a horizontal line. Then add another row of dots to the base. The number of dots you add is equal to the position of the figure in the sequence.

Lesson 4: Getting Loans

1. $258
2. B 25%
3.

End of year	Installment	Interest
1	$2,400	$192
2	$2,400	$144
3	$2,400	$96
4	$2,400	$48
5	$2,400	$0

4. **a.** $787.50; $141.75
   b. $11,151

Challenge Problem. Some key factors include asking customers for money up front, shortening the time that you give your customers to pay, lengthening the time before you pay your vendors, building up a savings fund.

Lesson 5: *Making the Numbers Work*

1. Sample answer: 24 ÷ 3 + 21 – (19 – 10)
2. Sample answer: (18 + 7) ÷ 5 × 2 + 10

Review

1. **a.** 110; **b.** $1
2. **a.** A $0.20
   b. $6 – ($2.00 + $\frac{1}{2}$ × $\frac{1}{5}$ × $2.00) = $3.80
3. Circle row 1.
4. C $960

Unit 7: Going on a Trip

Lesson 1: Budget Impacts

1. an increase in expenses of $306.50 per month.
2. A $399.54
3. Sample answer: about $650 to $700. Gas for car will be zero and utilities will be minimal. Rent, car payment, telephone, and cable TV will be the same. Food, clothes, and entertainment can be transferred to vacation costs.
4. **a.** Shade 3 squares; **b.** $36

Challenge Problem. ($200 – $46) ÷ $7; ($200 – $21.50) ÷ $8.50; 1 hour. You have to work 1 hour more.

Lesson 2: *Currency Translations*

1. B 20_8
2. Warren; Benton wrote 100_5 and translated that to 25_{10}.

Lesson 3: A Step to Your Future

1. 50%
2. **a.** 0
   b. 9 passenger should be marked on the number line.
3. B 22.5 minutes
4. Answers may vary but should add up to $6.95.

Challenge Problem. 4007. The value of the nth odd number is $2n – 1$, where n represents the count of that odd number. So for 1, n = 1; for 3, n = 2; for 5, n = 3; and so forth.

Lesson 4: Renting What You Need

1. $225; Sample: She didn't want the bother of carrying hers and the risk of possibly losing it.

Unit 7 *(continued)*

2. D $3,650
3.

Comparing choices	Fraction that shows the comparison	Amount of money that Beyonce saves
Choosing option i instead of option iii	$\frac{3}{4}$	$105
Choosing option i instead of option ii	$\frac{21}{26}$	$75
Choosing option ii instead of option iii	$\frac{13}{14}$	$30

4. a. $70
   b. Connect $70 and $0.

Challenge Problem. 21 girls and 18 boys

Lesson 5: Going Abroad

1. $198.90
2. $167 to $1,667
3. C a debit of $610.50
4. a. Use the number of pounds (2.205) to one kilogram and convert units
   b. Correct equation to (145 × $9.07) + (70 × $9.07) = $1,950.05.

Challenge Problem. 8. For example, use x, y, and z for the three numbers. Write equations: $\frac{x}{y} = \frac{3}{4}$; $\frac{y}{z} = \frac{2}{5}$; $x + y + z = 34$. Solve for y.

Review

1. ($635 + $152 + $500) ÷ 12. The effect is an increase in expenses of $107.25 per month.
2. Diagram shows that Arlan saves 80% of the cost of a repeat trip. An example diagram can be a circle graph in which the cost of the voucher, $200, is 80% of the cost of the $250 ticket.
3. Shade one quarter of the circle. [($106 + $18) – (46.50 × 2)] ÷ ($106 + $18) = $\frac{1}{4}$
4. B £307.28
5. 70 × $9.07 = $634.90

Unit 8: Keeping the Books

Lesson 1: Keeping Records

1. a. $5,108
   b. $1,617
   c. income, expenses, $3,491
2. $1,881
3. Answers will vary. A key factor is the difference between the cost of the project at $15 per hour ($1,537.50) and the offer of $1,200.
4. Advantages: Save money on interest payments. Simplify record keeping. Disadvantages: The rate may rise to a very high rate after 12 months. You may need a second card if you get close to exceeding the limit on one card.

Challenge Problem. 51. Some possible methods include test and check, write an equation, or others. For example, $10d + 5n = 7.85$, which simplifies to $2d + n = 157$; $d > n$; n is an odd number. If n were equal to d, then $3d$ would equal 157, and d might be about 53 and n would then be about 51.

Lesson 2: Club Records

1. underestimated; $124

Ski Trip Finances	Estimated cost	Actual cost	Variation
Hotel	$2,160	$2,135	$25
Transportation	$2,413	$2,567	–$154
Ski tickets	$2,100	$2,400	–$300
Food	$850	$545	$305

2. C 10
3. Shade sector A and label it 20%
4. ($238.00 – $117.60) ÷ $238.00 × 100% = 50.59%

Challenge Problem.

	24	15	2
9	3	3	1
10	2	5	1
8	4	1	2

Lesson 3: *Bingo Books*

1. Lance
2. C 6

Lesson 4: Big Events

1. $2,915; $1,700; make
2. a. (0.05 × $15 x 50) – (0.1 × $25 × 32)
   b. The event will make $42.50 less.
3. A Income: $3,360; expenses: $1,250
4. a. $21; b. $4,695

Challenge Problem. Key points to consider include the bookkeeping and managing the income and expenses for the event as a whole, coordinating dates and times, sharing responsibility, and reducing confusion.

Lesson 5: Spreadsheets & Software

1. a.

Total	$1,550	$1,450	$1,625	$4,625

   b. $4,625; does; These are two different ways of adding the same information.
2. C *Finances First*
3. 123 choose basketball; 77 choose softball; 64 choose robotics; 30 choose chess; 47 choose debate; 85 choose band

Challenge Problem. Some key advantages are speed, accuracy, and the ability to manipulate and present the data in different visual forms. Also, it is easy to add or change data and see the new results quickly. One disadvantage is the possible problem that can arise from rounding in translating numbers to percents and the other way around.

Review

1. $1,714 + $250 + $165 + $968 = $3,115; $1,092 + $623 = $1,715
2.

	Estimated cost	Actual cost	Variance
Hotel	$1,675	$1,654	$21
Transportation	$125	$134	($9)
Entrance tickets	$160	$140	$20
Food	$360	$382	($22)

   overestimated
3. a. $116 ÷ $358 × 100% = 32.4%
   b. Highlight $\frac{1}{3}$ of the line (from 66% to 100%).
4. does

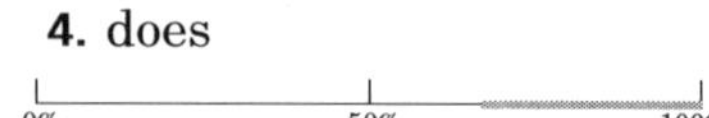

5. $2,150